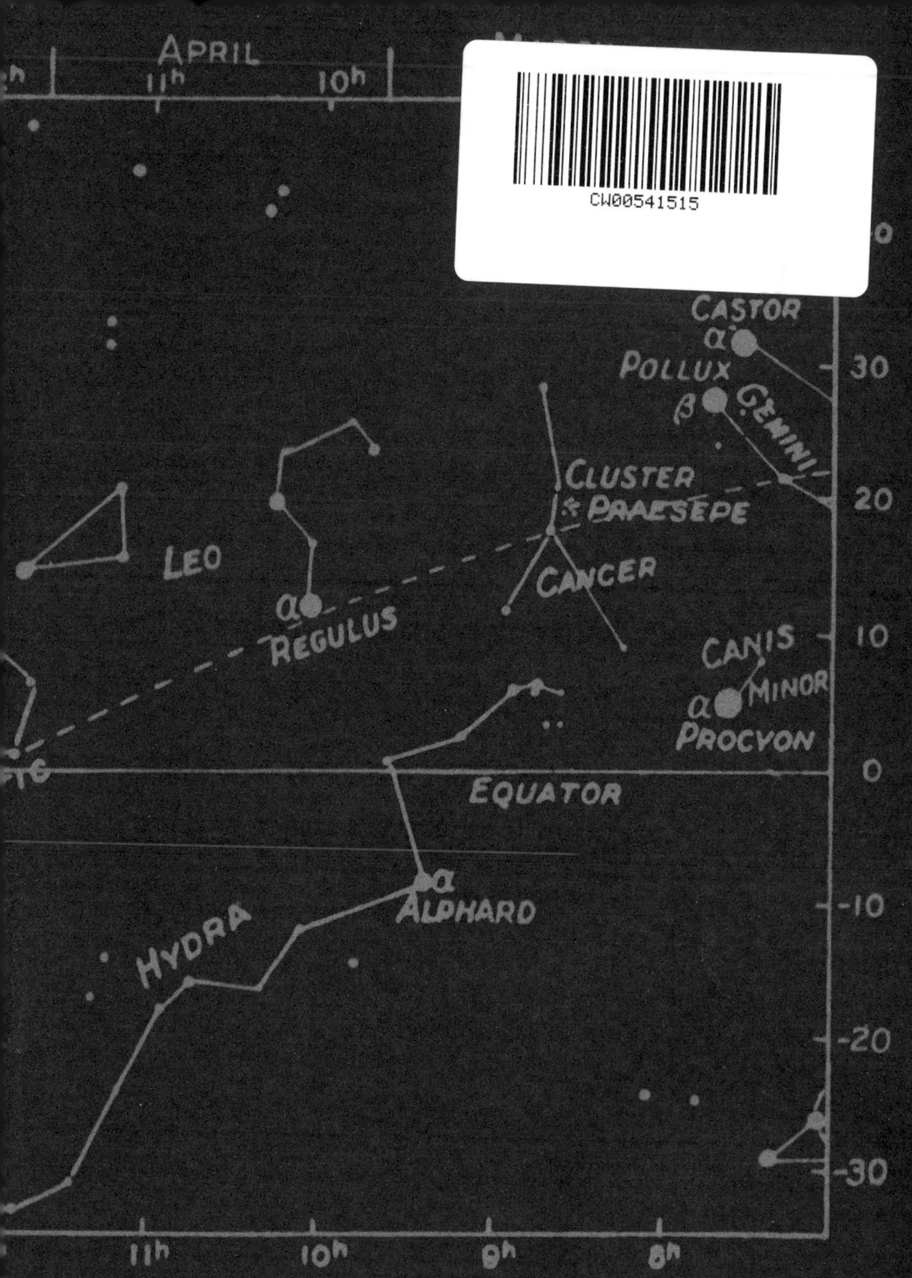
APRIL
11h
10h
CW00541515
CASTOR
α
POLLUX
β
GEMINI
30
CLUSTER
PRAESEPE
20
LEO
CANCER
α
REGULUS
CANIS
10
MINOR
α
PROCYON
0
EQUATOR
α
ALPHARD
HYDRA
-10
-20
-30
11h
10h
9h
8h

OUR WONDERFUL UNIVERSE

THE GREAT GLOBULAR CLUSTER IN THE CONSTELLATION HERCULES

It has been found by actual counting that there are upward of 50,000 suns in this cluster. Though they appear to be near together, neighbouring ones are in general one million million miles apart. This photograph was taken with the 60-inch telescope of the Mount Wilson Observatory, and the exposure was eleven hours, being continued on three nights.

Photograph by Ritchey, Mount Wilson Observatory

OUR WONDERFUL UNIVERSE

AN EASY INTRODUCTION
TO THE STUDY OF THE HEAVENS

by

CLARENCE AUGUSTUS CHANT
M.A. Ph.D. F.R.A.S. F.R.S.C
Professor of Astrophysics in the University of Toronto

ABOUT THE AUTHOR
CLARENCE AUGUSTUS CHANT

M.A. Ph.D. F.R.A.S. F.R.S.C

Professor of Astrophysics in the University of Toronto

Clarence Augustus Chant (May 31, 1865–November 18, 1956) was a Canadian astronomer and physicist.

He is considered by many to be the "father of Canadian astronomy", and indeed, five of his former students went on to become directors of astronomical observatories. Educated at the University of Toronto and Harvard, he taught at the University of Toronto from 1891 until his retirement in 1935. Chant was notable for his early work on X-ray photographs, but especially for his development of Canadian astronomy.

He organised the department at the University of Toronto and built up the Royal Astronomical Society of Canada (est. 1890) into one of the world's most successful organisations of its kind. In 1907, during his last year as President of the Royal Astronomical Society, he created the *Journal of the Royal Astronomical Society of Canada* and the *Observer's Handbook*. He would remain the editor of both publications until his death in 1956.

Chant participated in five total solar-eclipse expeditions, the most important being the one he led to Australia 1922 to test Einstein's theory of the deflection of starlight by a massive body.

In 1928 he published *Our Wonderful Universe* with enormous success; it was translated into five languages.

Through his efforts, the dream of a great observatory near Toronto came to fruition in 1933, when Mrs David Dunlap presented to the University of Toronto an observatory with a 74-inch (1.88 m) telescope. It remains to this day the largest optical telescope in Canada.

He died at 91 during the November 1956 lunar eclipse, while still residing at the Observatory House.

Asteroid 3341 is named in his honour, and in 1940, the Royal Astronomical Society of Canada created the Chant Medal, awarded each year to a Canadian amateur astronomer in recognition of their work in astronomy.

First Published 1928

by George G. Harrap & Co. Ltd.

39-41 Parker Street, Kingsway, London, W.C.2

Published in Great Britain in 2017 by

Papadakis Publisher

An imprint of Academy Editions Limited

Kimber Studio, Winterbourne, Berkshire, RG20 8AN, UK
info@papadakis.net | www.papadakis.net

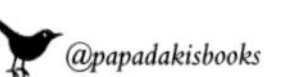

Publishing Director: Alexandra Papadakis
Design & editorial: Alexandra Papadakis
Cover design and design assistance: Aldo Sampieri
Production: Elizabeth Miller

ISBN 978 1 906506 62 9

A CIP catalogue record of this book is available from the British Library.
Printed and bound in China.

To

MRS DAVID A. DUNLAP

A GENEROUS FRIEND OF ASTRONOMY

THIS BOOK IS DEDICATED

PREFACE

HIS little book is designed to provide an easy and pleasant introduction to the study of the sky.

It is written in the form of a 'talk,' and spread over its pages are many illustrations, largely from actual photographs. The author hopes that young people, by reading the text and poring over the pictures, will be led to an intimate acquaintance with the heavenly bodies.

The subject is approached from the observational side. The reader is taken out in the open air, and his attention is directed to the various phenomena to be seen upon the dome of the sky which arches over him.

From the phenomena observed, by a simple and logical line of reasoning, the observer is led to realize the true meaning of the things he sees, and to understand why astronomers believe that the earth rotates upon an axis and at the same time revolves about the sun. It is shown how easily we can be deceived, and why we must exercise our reasoning power in order to decide what theories to accept or reject.

The development of our fundamental ideas regarding the general structure of the universe is unfolded in Part I. A certain effort of continuous thought is required to follow it, but the author thinks the effort is well worth while, and indeed is almost essential to a proper comprehension of the great universe. Moreover, it is quite within the power of the older pupils in primary schools, to say nothing of those farther on in their mental development.

Parts II and III deal, at some length, with the earth, the sun, the moon, the planets, the stars, and the other heavenly bodies. Any section of any chapter may be read by itself, and it will be found full of interesting facts.

The successive matters discussed are arranged in a definite connected sequence, so that the whole has a certain completeness about it; but the author wishes to say that this work has not been prepared with the intention that it should be used as a formal course of study. His chief object has been to give a clear and vivid picture of our great universe, so that his readers will actually *see* – with the eye of the mind – our family of planets revolving about the sun, and the myriads of celestial bodies, those other suns, existing far out in the depths of space. His aim is to excite the wonder of young people, to fire their imaginations, and to convey to them some notion of the majesty, the mystery, and the sublimity of it all.

Those who know nothing of the world of nature about them or of the heavens above them miss many of the intellectual and spiritual pleasures of life.

Above the doorways of many ancient Egyptian temples was carved a winged sun. This directed attention to, the beneficent Ruler of the Day, and also suggested the hope that those who entered might have their minds so illuminated that they would be able to comprehend the mysteries of life here and hereafter.

The author would humbly express the hope that all who seek to enter the Temple of Astronomy by way of the chapters which follow may receive mental enlightenment, and may also develop an increased reverence for the wonderful works of the great Ruler of the universe.

Let knowledge grow from more to more,
But more of reverence in us dwell.

C. A. C.

CONTENTS

PART I

THE CELESTIAL SPHERE AND ITS MOTIONS

CHAPTER I

THE CELESTIAL SPHERE

CHAPTER II

THE MOTION OF THE SUN AND THE MOON IN THE SKY

PART II

THE SUN AND ITS SYSTEM

CHAPTER III

THE PLANETARY SYSTEM – THE EARTH

CHAPTER IV

THE SUN AND THE MOON

CHAPTER V

MERCURY AND VENUS

CHAPTER VI

MARS

CHAPTER VII

JUPITER – SATURN – URANUS – NEPTUNE

CHAPTER VIII

ASTEROIDS – THE NEBULAR HYPOTHESIS – COMETS – METEORS

PART III
THE UNIVERSE OF STARS

CHAPTER IX

THE STARS IN THEIR SEASONS

CHAPTER X

THE NUMBER OF THE STARS; THEIR DISTANCES—THE NEBULAE

CHAPTER XI

DARK MARKINGS – CLUSTERS – THE NATURE OF THE STARS

ILLUSTRATIONS

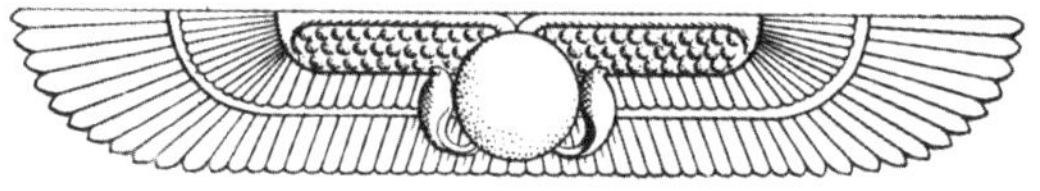

PART I

CELESTIAL SPHERE AND ITS MOTIONS

CHAPTER I

THE CELESTIAL SPHERE

LET us go out into the open air and look around us. We seem to be standing upon a level plain, at the centre of a great hemisphere formed by the overarching sky. It is not easy to get a perfect view of this hemisphere if one is in the midst of a city, since the houses obstruct our view; but when we are out in the open fields or on a wide expanse of water the full hemisphere of the sky is clearly seen.

One question naturally arises: Does the sky form only the half-sphere, or is there another half below our level plain and thus hidden from us?

As we continue in our study of the sky we shall be led to think that it really forms an entire sphere, one half being above our level plain, the other half below. This is shown in Fig. 1. Here we see a person at the centre of a level horizontal plain. Over him the sky forms a hemisphere, and under him, but unseen by him, it forms another hemisphere.

This sphere formed by the sky is called the celestial sphere.

The Daily Motion of the Sun

Suppose the time is nine o'clock in the morning. The sky is blue, perhaps with white clouds, and in the south-east we see a very bright object which we call the sun. It looks, up there among the clouds, like a round disc on the inner surface of the sky.

We know that it always rises in the east – although very few of us ever see it come up from beneath our horizontal plain! – that it

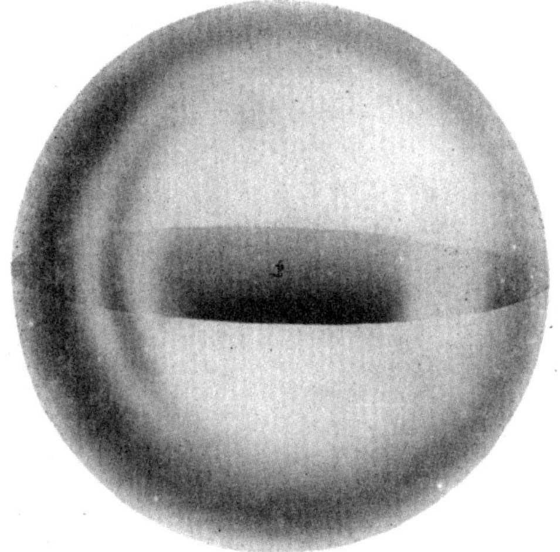

FIG. 1. THE CELESTIAL SPHERE
To the observer at the centre of the horizontal plain, with the sky forming a hemisphere over him.
Drawn by F. S. Smith

moves upward and westward, and that it reaches its highest position in the south at noon. We see it then continue to move to the west, and, gradually getting lower in the sky, we see it sink in the evening below the western horizon. In Fig. 2 we see the sun, first as its upper edge just appears above the eastern horizon, then in its position at noon, and then just as it is disappearing below the western horizon.

Here is another question. Does the sun move *along* the sky, or is it *fastened upon it*, so that the sun and the sky must move together? Did you ever think about that?

The Daily Motion of the Moon

At night the scene is changed. The sky is dark blue, almost black, in colour; and on it are numerous bright dots which we call stars. Some are quite brilliant, while many are faint and can be seen only

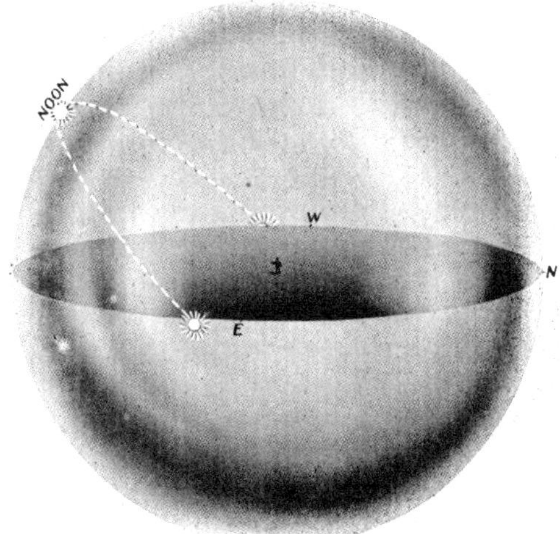

Fig. 2. The Celestial Sphere

To the observer at the centre the sun appears to rise in the east, to move upward and westward, reaching its highest point at noon, and then to move over and set in the west.

Drawn by F. S. Smith

by looking carefully for them. Groups of these stars seem to form triangles, squares, a plough, and other figures. We shall examine some of them more closely a little later.

Probably the moon can also be seen in one of its many shapes, or 'phases,' as they are called. Suppose it is round like the sun – in which case it is said to be 'full' – and we see it over in the east, near the horizon. Let us watch its behaviour. We find that it continually moves over to the west, and disappears below the western horizon just as the sun does.

Indeed, whatever the moon's shape may be, if we closely observe it we shall see it move over and set in the west.

The Daily Motion of the Stars

Let us watch the stars also. Perhaps there are three of them forming a triangle low down in the east, and several arranged in a straight line in the west. In the north we see the seven stars known in England as the Plough, in America as the Dipper. They form a part of the Great Bear constellation. On an autumn evening the Plough appears right side up.

An hour or two later let us look at these stars again. Those in the east are much higher, those in the west are much lower – or perhaps have disappeared altogether – while the Plough, or Dipper, has turned so as to rest partly upon its handle.

FIG. 3. A SCENE IN QUEEN'S PARK, TORONTO – DAYLIGHT VIEW
The photograph was taken in winter, from a window in the Provincial Parliament Buildings, looking north-west.

It obviously looks as though the entire sky, carrying the sun, the moon, and all the stars, is in motion, turning steadily from east to west, and making a complete revolution in one day!

Photographs by Day and by Night

Now a good way to study the motion of a bright object is by photography. Here is a photograph of a familiar scene in Toronto, taken on a winter afternoon (Fig. 3). The camera was placed on the window-sill in an upper story of the Provincial Parliament Buildings, facing the north-west. Through the leaf-less oak trees you can see Hart House, of the University of Toronto, and just to the right of it is one of the small domes of Trinity College. There are two motor-cars coming south, while three persons are walking over the snow. As the motors seem to be standing still the exposure must have been only a very short one – a small fraction of a second.

Next let us look at a photograph (Fig. 4) which was taken at night with the same camera in the same position, the exposure being one of five minutes. You see the electric lights along the road and also the lighted windows of the buildings. The road itself appears bright, but where are the cars? As a matter of fact, there were many of them coming along the road. The headlights of each car as it moved forward made two curved streaks of light on the plate, and there were so many cars and so many bright streaks that at last the whole road seemed to be lighted. If you look closely you will see the lights of a single car as it turned outward to pass a car ahead of it. Also the trails from two or three cars can be seen as they turned to their left to go eastward.

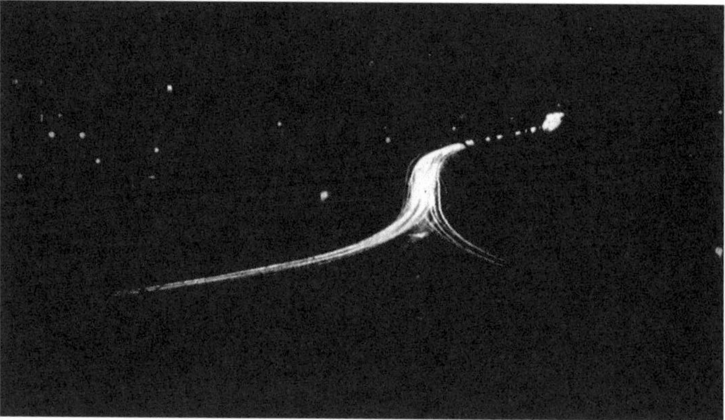

Fig. 4. A scene in Queen's Park, Toronto – Night View
For this photograph the camera was placed, at 8 P.M., in the same position as for the photograph shown in Fig. 3, but the exposure was one of five minutes.

After this, with the camera in the same position, a picture was taken with an exposure of twenty minutes, and here it is (Fig. 5). You can now see the trees and the buildings. The many cars which have come along have left trails of light which show the roads very clearly. Notice also the trails of several cars which turned outward to pass others, and the trail of a car which came from the left and went directly eastward.

FIG. 5. A SCENE IN QUEEN'S PARK, TORONTO – NIGHT VIEW, LONGER EXPOSURE
For this photograph the exposure was twenty minutes, and much more detail is shown.

Photographs of the Stars

Now let us experiment with the stars.

If you face the north and raise your eyes about forty-five degrees or a little more (depending on your latitude) you see the Pole Star, with other stars about it. In Fig. 6 is a map of the stars as they are seen about 9 P.M. on November 1. The Pole Star is at the centre. It is at the end of the tail of the Little Bear. In America this constellation is often called the Little Dipper, though it is not so well shaped as the other one.

Next we shall try to photograph them. Choosing a place where the sky to the north is not very bright, we mount the camera on a window-sill or other solid base, and tilt it upward so that it is directed toward the Pole Star. Having focused the camera for a great distance, we open the shutter and expose for a long time – several hours if possible. Fig. 7 shows the kind of picture we get. It was taken from a city window with an ordinary camera, the exposure being one of about two hours. You see a large number of trails, each one being produced by a star. You notice, too, that they are all arcs, or portions, of circles having a common centre.

From our photograph we are led to think that the stars in the north are moving in circles round a common centre, and we should expect the length of the trails they make to be proportional to the length of the exposure. The heavy trail near the centre was produced by the Pole Star. Notice that this star is not exactly at the centre of the circles, though it is near to that point.

Of course, somewhat better pictures can be taken with a camera made specially for sky photographs. Here is a photograph taken at the

FIG. 6. THE POLE STAR WITH THE STARS ABOUT IT
This chart includes stars within 30° of the pole of the sky. The Pole Star is at the centre, and the brightest star seen in the bottom is one of the pointers in the side of the Plough or Dipper.

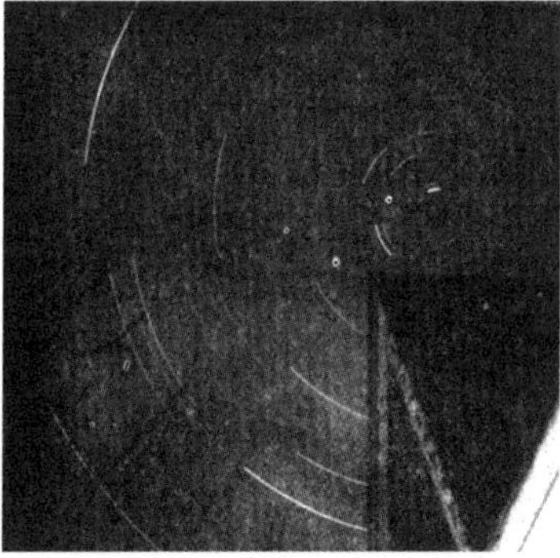

FIG. 7. CIRCUMPOLAR STARS. AMATEUR PHOTOGRAPH
This photograph was taken with an ordinary camera, which was placed on a window-sill. Exposure, two hours. The centre of the circles is the north celestial pole. The bright trail near it was made by the Pole Star.

Lick Observatory in California (Fig. 8). As you see, the exposure was a long one. How long was it? At what season do you think the picture was taken?

Returning to our own photographic experiments, next turn the camera to the south, tilt it up about 45°, or a little less, and make another exposure. An hour is long enough this time. This is what we get (Fig. 9). The camera with which this picture was taken was a very ordinary one. Indeed, any person can take such pictures. Try the experiment.

All the Stars describe Circles

Let us look closely into this picture.

Near the middle are three parallel trails which seem to be quite straight. These were made by the three stars in the Belt of Orion. As perhaps you know, Orion is one of the glorious winter constellations. Above is a trail curving upward. It was made by the famous red star named Betelgeuse. A short distance below the centre is a bright trail which was made by the

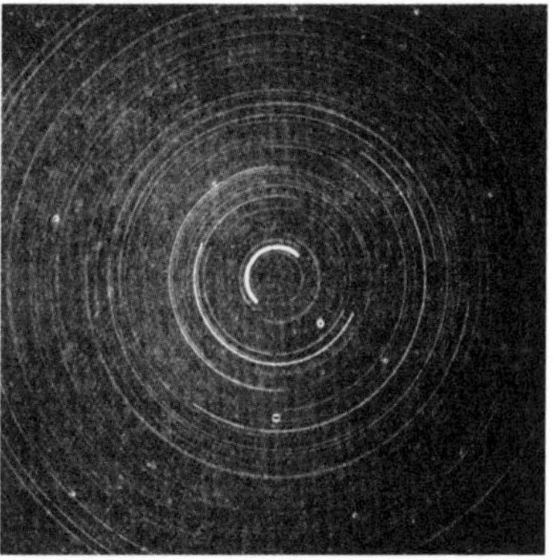

FIG. 8. CIRCUMPOLAR STARS
For this picture an exposure of twelve hours was give on a long winter night
Lick Observatory photograph

star Rigel, while near the bottom is the trail made by Sirius, the Dog Star, which is the brightest star in the sky and hence makes a very bright trail. Note that the trails of Rigel and Sirius curve downward. These trails, as well as that of Betelgeuse, are arcs of circles. As nearly as we can see, then, the stars in the Belt describe straight lines while the others describe circles.

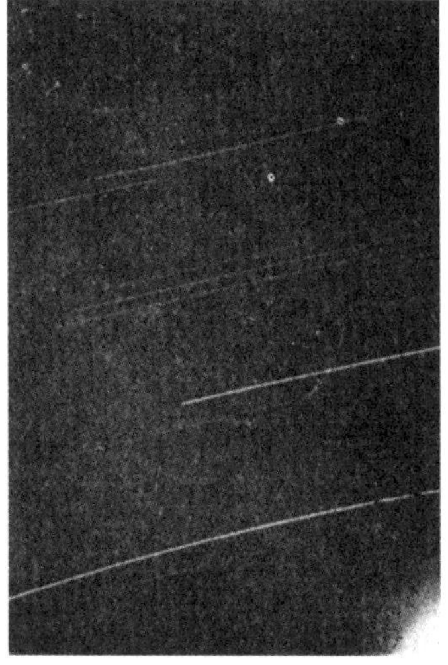

FIG. 9. EQUATORIAL STARS
AMATEUR PHOTOGRAPH
For this picture the camera (an ordinary one) was tilted upward about 45°, facing south. Exposure, one hour. The three trails at the centre were made by the stars in the Belt of Orion.

Now how can we explain all this?

It is quite easy. The entire sphere formed by the sky, together with the stars, which seem to be fastened upon it, appears to turn about an axis, and thus every star describes a circle. The axis passes through the common centre of the circles which we saw in the photographs of the polar stars (Figs. 7 and 8) and also through the common centre of the circles described by the stars in the Southern Hemisphere, which we cannot see.

The point in the north is called the north pole of the sky, or the north celestial pole; that at the south is the south pole of the sky, or the south celestial pole. These are shown in the diagram (Fig. 10).

The Pole Star is near, but not exactly at, the north celestial pole; there is no star near the south celestial pole. Midway between these poles is the equator of the sky, or the celestial equator.

From our observations and experiments, then, it appears that the sun, the moon, and the stars are attached on the inner surface of the celestial sphere, which turns about an axis, carrying all these bodies with it and making a complete rotation in one day. This would explain why the sun, moon, and stars rise in the east, cross the sky, and set in the west.

Does the Sky really move?

But beware now! Stop and think!

Is it really the fact that the sun, the moon, and all the stars in the sky revolve about the earth once a day? They certainly appear to do so, and the people of ancient times believed that they actually did. Are we deceived? Is there any other way to account for what we saw? Yes, there is.

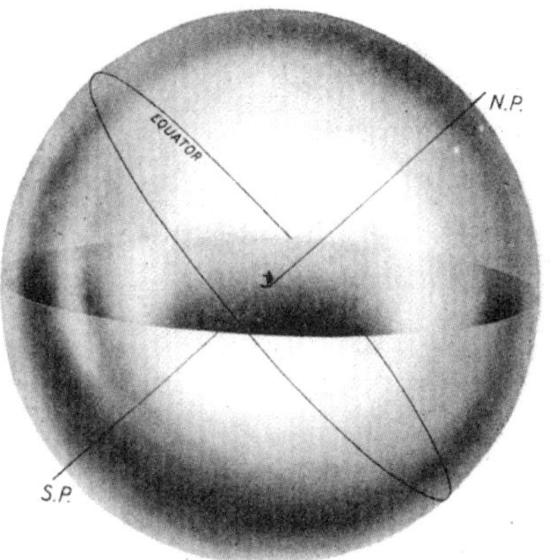

FIG. 10. THE CELESTIAL SPHERE, SHOWING THE NORTH AND SOUTH CELESTIAL POLES
The sky appears to rotate about an axis which passes through the celestial poles, and the celestial equator is half-way between them.
Drawn by F. S. Smith

All this time we have been assuming that the earth on which we live is fixed in place and at rest; but how would the sun and the stars appear to behave if *we* were in motion while *they* were at rest?

Suppose you wish to take a railway journey. You go down to the station and take your seat in the passenger train. There is another train near by, and as you look out of your window you see it begin to move – or at least appear to do so. For some time you may not be sure whether it is your train or the other one which is moving, and you may perhaps have to look at the wheels of the other train to see if they are turning. It looks as if the other train was going backward, but you find out that your train is going forward while the other is standing still.

It is just the same with the earth and the stars. It looks to us as if the sky, carrying the stars, is moving from east to west; but things would look just the same if the stars were standing still and the earth was turning in the opposite direction – that is, from west to east.

FIG. 11. A CAMERA AND FLASHLAMP ARRANGED FOR EXPERIMENTS
The flashlamp may be rotated while the camera is at rest, or the camera may be rotated while the flashlamp is at rest.

How are we going to tell whether it is the earth or the sky which moves?

Experiment with a Camera and a Flashlamp

In Fig. 11 is shown an apparatus with which we can make some interesting experiments. At the left-hand end of a board is a flashlamp (A) mounted on a wooden arm (B), which can be made to revolve by turning the crank C. Thus, if the crank is turned the bright little light of the flashlamp will describe a circle. On the other end of the board is mounted a camera, which also can be rotated – by the crank D.

The following experiment can be performed. The apparatus is taken into a darkened room, and, first of all, while the camera is at rest the crank C is turned, making the flash lamp describe a circle. You would expect the picture made by the camera to be simply a bright circle. Next the flashlamp is kept fixed, and the camera is rotated by turning the crank D.

What is the result? The two pictures are shown in Fig. 12. Each is a circle, and you cannot distinguish one from the other!

So it is with the stars and the earth. You cannot say whether the stars actually describe circles or the camera rotates as it is carried by the earth.

How can we find out, then, which moves? There are several experiments which have been devised to test whether the earth rotates, but they are all rather difficult to perform. Perhaps the best known is that in which a special kind of pendulum is used; but the

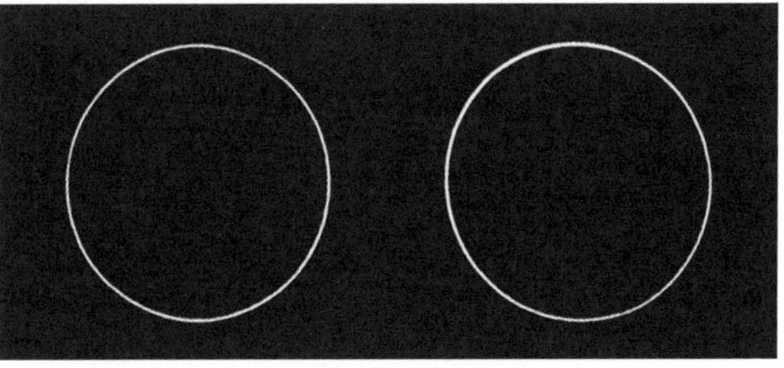

FIG. 12. RESULTS OF THE EXPERIMENTS: TWO CIRCLES EXACTLY ALIKE
One circle was obtained by rotating the flashlamp, the other by rotating the camera, shown in the last picture.

gyroscope and the experiment in which bodies are dropped from a great height also have been tried. All these experiments are described in more advanced books on astronomy.

The Gyro-compass shows that the Earth rotates

One of the most useful things on an ocean-going ship is the compass. Without it the captain could not find his way to whatever port he wishes to make. In recent years the ordinary magnetic compass has been to some extent displaced by what is known as the gyro-compass. Indeed, in a submarine this is the only kind of compass which can be used, since a magnetic compass is useless if it is completely surrounded by iron.

A photograph of the mechanism of a gyro-compass is shown in Fig. 13, while the compass in its case on board ship is shown in Fig. 14. The chief part of it is a heavy wheel which can be made to rotate very rapidly. Of course, it has to be very carefully balanced, or the apparatus would be quickly shaken to pieces. The wheel is enclosed in the case (A), and in the type of instrument shown in the picture it is 12 inches in diameter, weighs 45 pounds, and rotates 8600 times a minute. In some other types the number of rotations per minute is nearly twice as great, but the wheels are not so large and heavy.

The wheel and its case are mounted in a very delicate and accurate way. Now this wheel, as it spins about, always moves until its axis is in the north-south direction – until the axis is in the meridian plane, as the astronomer says. The graduated circle above is

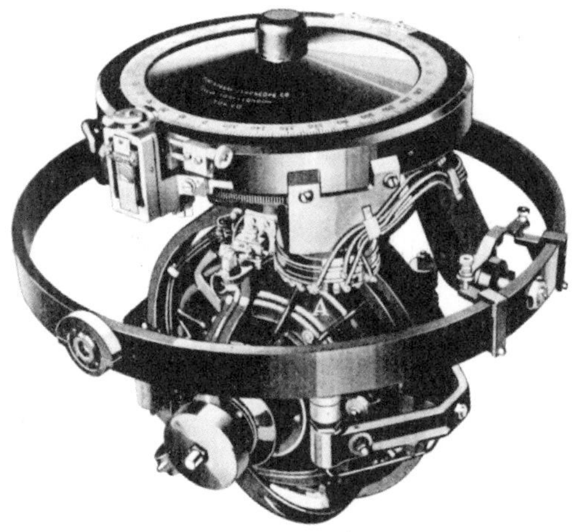

FIG. 13. THE MECHANISM OF A GYRO-COMPASS
This instrument indicates true north and south, and it would not do this if the earth did not rotate.
Photograph from the Sperry Gyroscope Co.

FIG. 14. A GYRO-COMPASS IN ITS BINNACLE
The gyro-compass in its protecting case is in a place convenient for the navigating officer to read it. Sometimes the main compass controls several auxiliary dials in other parts of the ship.
Photograph from the Sperry Gyroscope Co.

connected with the case, and from it the officer in charge can learn exactly in what direction the ship is heading at any time.

Now if the earth did not rotate on its axis the gyro-compass would not move about and set itself as it does; and so we have good reason to believe that it is the earth which turns, not the stars. The axis about which the earth turns is the line about which the celestial sphere *seems* to turn.

Suppose you were standing at the North Pole of the earth, where would the north celestial pole be? Right overhead.

The Cause of Day and Night

Thus we are convinced that the earth is rotating on its axis, turning from west to east. It is this motion which causes day and night.

Suppose it is night-time. The sky is dark – there is no sun to give us light. The earth continues to turn on its axis, until at last we see the sky becoming rosy in the east. Then the sun itself appears to be rising at the eastern horizon. It is the earth which has turned about and brought us so that we can see the sun.

Then as the earth turns farther the sun appears to get higher in the sky, until it reaches our meridian and it is noon. The earth continues to turn from west to east, and the sun appears to move westward, getting lower and lower in the sky until at last it disappears below the western horizon. We say that the sun has set.

We know, however, that the people who live west of us on the earth are still enjoying the sun's light and heat, while for those on the very opposite side of the earth from us the sun is rising. In the words of the well-known hymn,

> The sun, that bids us rest, is waking
> Our brethren 'neath the western sky.

CHAPTER II

THE MOTION OF THE SUN AND THE MOON IN THE SKY

The Path of the Moon among the Stars

OW let us give some attention to the moon. As we have already observed, it rises and sets like the sun; but it will be interesting to find out if it stays in the same place in the sky while the latter is turning round each day – or, as it is more accurate to speak so, while the earth is rotating on its axis.

We can easily do this by watching where the moon is among the stars night after night.

You know that you can see the moon and the stars at the same time, and so we will make a map of the stars and carefully mark the position of the moon on it. Here (Fig. 15) is a chart of the sky which shows the position of the moon and its shape on the evening of February 20, 1926, and on the six following evenings. On the 20th it was near Aldebaran, the brightest star in the constellation Taurus (the Bull), and from the chart you can see how it moved through this constellation and then through Gemini (the Twins) and Cancer (the Crab).

By continuing to chart the moon night after night we find that all the time it is moving eastward, and that at the end of a month it has moved completely round the sky and has come back to the place among the stars where it began.

Let us stop a minute to ask if we are not deceived again. Does the moon actually revolve about the earth? Yes, it does; and the time it requires to go completely round is called a *moon*-th or, as we say, a *month*.

The Motion of the Sun in the Sky

Having settled the motion of the moon, let us investigate the sun.

We wish to know if the sun stays at the same fixed place on the sky as though it were nailed there.

In this case we are confronted with a difficulty, since the sun is so bright that we cannot see it and the stars at the same time, and therefore cannot draw a map to show where it is among the stars day after day. We might even wonder if there are stars in the sky around the sun at all.

However, the moon sometimes comes directly in front of the sun and shuts off its light. We say that the sun is totally eclipsed. At such a time the stars about the sun can be seen and photographed. Also, if in the daytime a telescope is pointed in exactly the right direction, one can see the stars. Thus, we are sure that there are stars in all parts of the sky all the time, but, as has been said above, the brightness of the sun is so great that we cannot see just where it is among them.

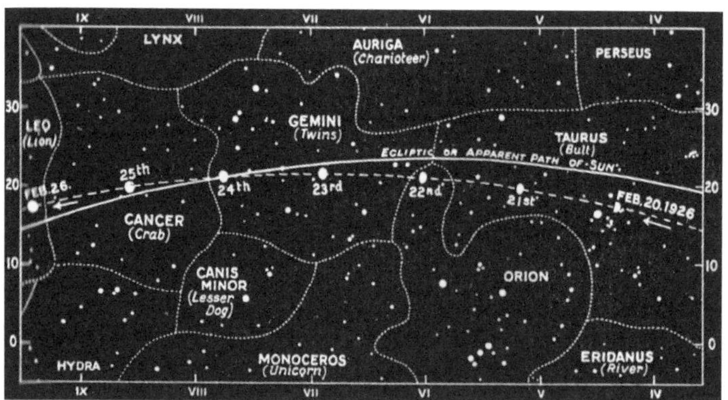

Fig. 15. The Path of the Moon among the Stars
February 20-26, 1926
During this interval the moon passed through the constellations Taurus, Gemini, and Cancer, and also changed its shape.

Now we know that the sun moves north and south in the sky. In the winter it is much farther south than in the summer. This is well illustrated in Fig. 16. In the centre is Stonehenge, perhaps the most famous of ancient English monuments. It stands on Salisbury Plain, in Wiltshire, and, it has been estimated, was constructed about 3600 years ago. Many believe that it was used in the performance of religious ceremonies.

There are many stones, some of great size, arranged in concentric circles. If you stand on the large, flat stone at the centre called the altar-stone, and look out between two great upright stones and over the top of another upright stone called the Friar's Heel some distance away (marked A in the picture), you will face precisely that point on the horizon in the north-east where the sun rises on June 21, the longest

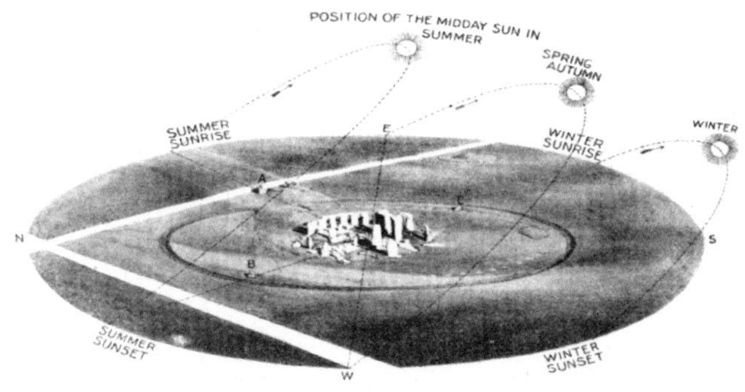

FIG. 16. STONEHENGE, SHOWING CHANGE IN THE HEIGHT OF THE SUN WITH THE SEASONS
The stones were placed so as to indicate the points where the sun rises and sets on June 21 and December 21. The sun at noon is much higher in summer than in winter. The air-photographs from which this drawing was made were obtained by courtesy of Sir Charles Close.
Drawn by F. S. Smith from air-photographs.

day of the year. Fig. 17, drawn from photographs, shows a company of people gathered within the circle of the stones on the morning of June 21 in order to see the sun rise behind the Friar's Heel. Another stone (marked B in Fig. 16), to the north-west, shows the direction to look to see where the sun sets on that same date. Again, if you look from the altar-stone toward a stone (C) in the south-east direction you will face that point of the horizon where the sun rises on December 21, the shortest day in winter. It seems likely that the stones were placed in these positions in order to show the place of the sun's rising on the dates mentioned. The picture certainly shows how much higher the sun is at noon in summer than in winter.

But does the sun also move eastward among the stars, as the moon does? Yes, it does. The astronomer has found a way (which we cannot explain here) to locate its position among the stars

FIG. 17. VISITORS TO STONEHENGE WATCHING THE SUN RISE ON JUNE 21
Many people gather within the circle of stones to observe the sun as it rises from behind the Friar's Heel on June 21.
Drawn by F. S. Smith from photographs by Mullins and others

from day to day. The path of the sun among the stars from May 15 to July 14 is shown in Fig. 18. During these two months it moves through the constellations Taurus and Gemini, and, of course, we cannot see the stars in those constellations during these months.

The Path of the Sun (the Ecliptic)

Let us follow the course of the sun through the seasons. On March 21 it is on the celestial equator, and the days and nights are equal. Then it moves north and always east, daily getting higher in the sky at noon, until June 21, when it is farthest north of the celestial equator

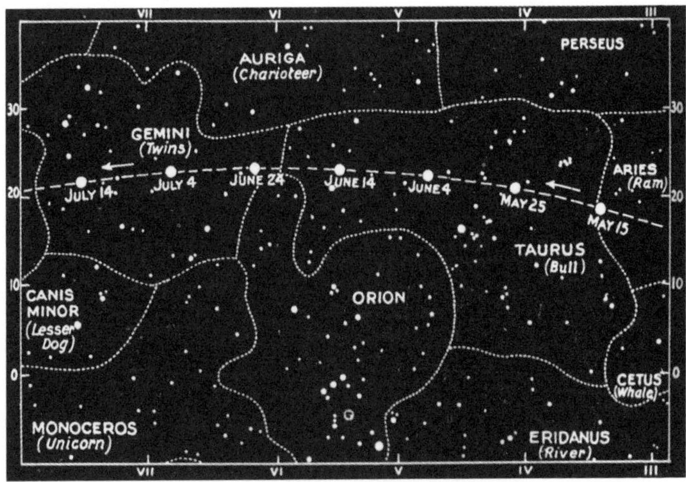

Fig. 18. The Path of the Sun among the Stars
May 15 to July 15
Every year during May, June, and July the sun follows the same path through Taurus and Gemini. The apparent path of the sun is the ecliptic.

and the daylight is longest. This is the summer solstice. Then it turns and moves south, and, always continuing eastward, it reaches the celestial equator again on September 22. Then it goes on farther south, and, of course, always east, until December 21, when the daylight is shortest. This is the winter solstice. It now turns about and moves north, and, still going east, in the course of three months it reaches the celestial equator again, on March 21. This is the spring equinox, and once more the days and nights are equal.

Thus half of the sun's path is north of the equator, and half is south.

The path which the sun follows in the sky is called the ecliptic, and year after year the sun travels over exactly the same track. The length of the year is the exact time taken by the sun to go completely round the ecliptic. This path of the sun, the ecliptic, is shown in Fig. 19.

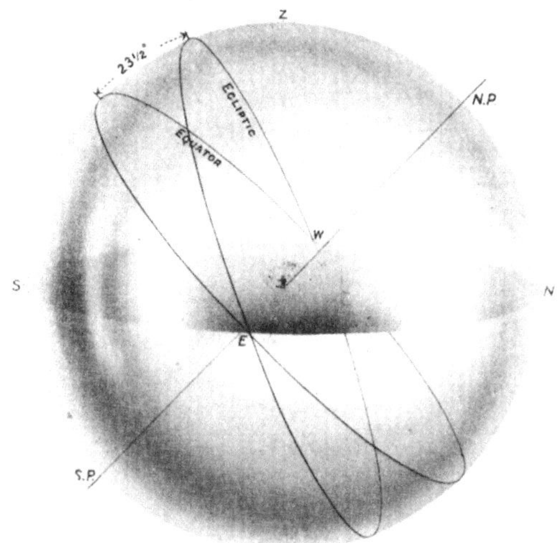

FIG. 19. THE CELESTIAL SPHERE, SHOWING CELESTIAL EQUATOR AND ECLIPTIC
The celestial equator is midway between the celestial poles, and the ecliptic, which is the sun's apparent path in the sky during the year, is inclined to it at an angle of 23½°. *Drawn by F. S. Smith*

If only some kind fairy would fly up to the sky and mark on its surface the path followed by the sun, and also the celestial equator, it would be a great convenience to people studying astronomy! It cannot be done, however; and you must stir up your imagination, and, with the eye of the mind, see the great sun up there in the sky ceaselessly moving forward day after day and year after year along its appointed path.

The ancient Greeks represented Phœbus, the sun-god, as driving the chariot of the sun ever onward among the stars. His roadway was the ecliptic, and with great skill he always drove his car exactly along it, never swerving to right or to left. It is related that on one occasion his son Phæthon begged to be allowed to drive the car, and, though Phœbus hesitated, he at last consented. The reckless youth drove carelessly, got off the road, and nearly burned up the earth. Fig. 20 is taken from an old book printed at Venice in 1482.

It represents Sol (the Sun) in his chariot driving his fiery horses. Surely now you can picture to yourself the mighty sun moving along its path on the celestial sphere.

FIG. 20. SOL (THE SUN) DRIVING HIS CHARIOT

This figure of Sol driving his chariot drawn by four fiery horses is from a book printed at Venice in 1482 by a noted printer named Ratdolt. It was one of the first books illustrated with woodcuts, among them being representations of the constellations – the earliest known. The name of the book is *Poeticon Astronomicon* (*Astronomy in Verse*), and its author was Hyginus, who was librarian to the Roman Emperor Augustus (63 B.C. – A.D.14). It was written in Greek.

By courtesy of the Librarian of the U.S. Naval U.S. Naval Observatory

The Zodiac

Imagine a long ribbon 16° wide (that is, thirty-two times the diameter of the sun) to be tacked on the celestial sphere so that the ecliptic is precisely along the middle of it. This is the zodiac. Within that belt of the sky the moon and the planets are always to be found. From the earliest times the zodiac has been considered to be divided into twelve equal portions, each being 30° long and 16° wide. These are called signs of the zodiac. Their names and the symbols used to represent them are Aries (♈), Taurus (♉), Gemini (♊), Cancer (♋), Leo (♌), Virgo (♍), Libra (♎), Scorpio (♏), Sagittarius (♐), Capricornus (♑), Aquarius (♒), Pisces (♓).

With a little study, using the star maps in Figs. 103, 105, 107, and 109, one can learn the constellations which are along the ecliptic, and thus be able to locate its position in the sky at any time that the stars are visible.

You must remember that the annual motion of the sun round the ecliptic is quite different from its apparent daily motion, in which it seems to rise in the east and set in the west, giving us day and night.

The Sun's Motion Apparent, not Real

But, after all, does the sun really travel round the sky in the course of a year?

No, it does not! We have been deceived again! It is the earth which actually moves about the sun, though it *seems* to us that the sun is moving about the earth.

In Fig. 21 you will see how this happens. The earth actually travels in an oval, or rather elliptical, path about the sun, as shown by the ellipse in the diagram. When, on January 1, it is at A the sun appears to be at *a* in the ecliptic, among the stars in the constellation Scorpio (the Scorpion). Consequently the stars in Scorpio and in other constellations in the same part of the sky, such as Corona (the Crown) and Hercules, cannot be seen at that time; but when at that season the earth, by rotating on its axis, has brought us night we look in the opposite part of the sky from where the sun is, and there see Taurus, Orion, Perseus, and other winter constellations.

Three months later, on April 1, the earth is at B, and the sun appears to be at *b*. Consequently we cannot see Cygnus (the Swan), Pegasus (the Winged Horse), Aquila (the Eagle), since they are in the same part of the sky as the sun; but we can see Leo (the Lion), Ursa Major (the Great Bear), and other constellations which are in the opposite part of the sky. On July 1 the earth is at C, and the sun appears to be at *c*; on October 1 the earth is at D and the sun appears to be at *d*; and then on January 1 the earth is back at A, and the sun 1 appears to be at *a* again.

Thus, while the earth actually moves in a little oval curve about the sun, the sun appears to describe a great circle in the sky – the ecliptic, as you have already learned that this circle is called.

The Stars change with the Seasons

Fig. 21 shows why we see different stars at different seasons, but, as it is very desirable that you should understand this clearly, we shall consider the matter a little longer.

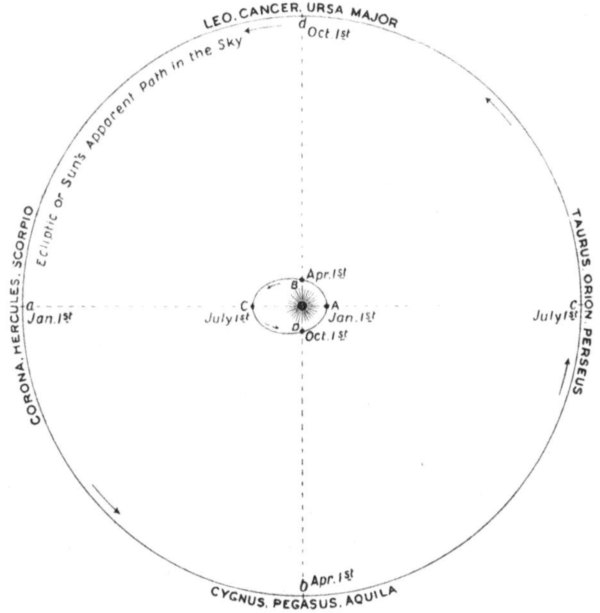

FIG. 21 THE SUN'S APPARENT PATH AND THE EARTH'S REAL ORBIT
Actually the earth moves about the sun in a path which is elliptical in shape, the sun being not at the centre but in a focus of the ellipse. To an inhabitant of the earth who does not realise that it is in motion the sun appears to move about the sky in a great circle which we call the ecliptic.

At night the sky looks like a great, dark, spherical shell with the stars upon its inner surface. In Fig. 22 is shown an attempt to represent the appearance of such a shell as seen by a person viewing it from the outside.

Notice that the stars are distributed over the entire surface, bright ones and faint ones being seen in every quarter.

At the centre is the sun, and about it revolves the earth, keeping its axis always pointed in the same direction. When in position *a* it

FIG. 22. THE STARS AND THE SEASONS
At night the stars appear to be on the inner surface of a great dark sphere. This picture is intended to show such a sphere, made of transparent material and seen from the outside. The sun is at its centre, with the earth revolving about it.
Drawn by F. S. Smith

is winter in the northern hemisphere of the earth, since the sun's rays fall short of the North Pole, but reach beyond the South Pole.

Now remember that you are upon the earth, and, let us say, in the northern hemisphere. In the daytime, of course, you will see the sun in the sky. Then, by the rotation of the earth about its axis, night comes, and as you look around what stars will you see? Clearly those in that part of the sky marked A in the picture. The sun will appear to you to be in that portion marked C, and the stars there will not be visible to you.

Three months later the earth will be at *b*, and it will be spring-time in the northern hemisphere. The stars seen during the spring nights are those on the farther side of the sphere, on the opposite side from the sun.

In another three months the earth will be at *c*, and summer will have come in the northern hemisphere. During the summer nights you will look out upon the stars in that part of the sky about C, while those about A, being in the direction of the sun, will not be visible.

Then three months later, during the autumn nights, the stars to be seen will be those in that part of the sphere which is nearest to the person viewing it from the outside.

Thus the stars change with the seasons, though they are the same at the same season year after year.

Next consider those stars at the upper part of the picture toward which the axis of the earth points – in other words, those about the north celestial pole. At all seasons a person in the northern hemisphere will be able to see them, while a person in the southern hemisphere will be able to see the stars about the south celestial pole.

A View of the Universe

So we have been deceived twice, and we should learn the lesson that things are not always what they seem.

Why were we deceived?

It was because the earth moves so gently, without jolt or jar, as it rotates on its axis and at the same time moves along its path in space about the sun, that we do not feel that we are in motion at all.

If we could only travel far up in space, entirely away from the earth and the sun, then we should see things as they really are. We should then have an aeroplane, or bird's-eye, view of the universe. Fig. 23 shows what we should see. Here are two little celestial travellers who have been transported far up in space toward the Pole Star, a very great distance – perhaps ten thousand million miles – and allowed to see our wonderful universe from that place.

And what do they see?

Far down below them, in the direction from which they came, they see the solar system. There is the sun, still very bright, even though they are so far away from it. Then, as they gaze steadily, they see a number of bodies moving about the sun. These are the planets. The travellers notice that each of them looks like a bright, semicircular disc, and wonder why. They soon decide that the planets must be spherical in shape, that they are really dark bodies, and that only that hemisphere which is turned toward the sun is lighted up an thus made visible.

The travellers watch the planets closely, trying to count how many there are and to observe how fast they move.

Nearest the sun is Mercury. It is the smallest of all, and speeds along most rapidly. Next is Venus, much larger than Mercury but moving more slowly. The next is the earth, of about the same size as Venus and travelling still more slowly. Our travellers observe, also, that the earth is accompanied by a round body which continually revolves about it. That is our moon. The fourth planet is Mars. It has a diameter only about one half that of the earth, and it is accompanied by two tiny moons.

Then there is a wide gap, and the fifth planet is Jupiter, by far the largest of all. It has four large moons, and several small ones which they can hardly see. The sixth member of the family is Saturn. It also is large, though much smaller than Jupiter. It possesses a wonderful set of rings and a company of moons. After this comes the seventh member, which is named Uranus, with four moons; and then following the eighth and last, Neptune, with one moon, travelling along in its distant path – like a lone sheep with a single lamb! Those planets which are far out from the sun move in their orbits more slowly than do the planets nearer the sun.

FIG. 23. A BIRD'S-EYE VIEW OF THE UNIVERSE
These little celestial travellers have wandered far, far off in space, toward the Pole Star, and are surveying the universe. They see our bright sun and the family of planets revolving about it. They also observe two comets which have come out of space to visit the sun. The stars in their constellations, however, look just the same as they did from the earth.
Drawn by Henrietta N. Hopper

The travellers see also two comets which have come out of space to visit the sun. They approach it, pass round it, and then move off, probably never to return.

What could be more thrilling than to see this great family of planets move majestically along in their orbits while their attendant moons continually revolve about them as if to protect them from any danger in the way?

Turning their eyes from the family of the sun, the travellers survey the other objects in the sky. In every direction they see the stars standing out there perfectly still, like great lamps fixed out in the depths of space. To their surprise the travellers recognise the very same constellations which they were familiar with while upon the earth, and conclude that the stars must be at immense distances.

Far over there is Orion and his wonderful belt, the Dog Star, the Bull, the Milky Way, and all the rest!

It is a wonderful universe!

PART II

THE SUN AND ITS SYSTEM

CHAPTER III

THE PLANETARY SYSTEM – THE EARTH

A Look at the Planetary System

LET us now look more closely at some of the things we have seen and learn something further about them.

First of all let us consider the relative sizes of the orbits of the eight planets. They are shown in Fig. 24. Notice that there are four quite close to the sun. These are the orbits of Mercury, Venus, the earth, and Mars. Then there is a wide space, and beyond this are the orbits of Jupiter, Saturn, Uranus, and Neptune. Thus we can divide the planets into two groups – those near the sun and those far away.

In the wide zone between the orbits of Mars and Jupiter small bodies called planetoids or asteroids are to be found. A large number have been discovered, the majority of them being visible only through a great telescope.

Let us look at each group separately. Here are the orbits of the four inner planets (Fig. 25). Their distances from the sun are approximately 36, 67, 93, and 142 millions of miles. Perhaps you can remember these numbers. The dots show the position of the planets in their orbits every ten days. The place of each planet on January 1, 1927, is also shown.

Next we have the orbits of the outer planets (Fig. 26). Their distances from the sun are approximately 480, 880, 1780, and 2800 millions of miles. In the case of the orbits of Jupiter and Saturn the dots show the positions of the planets at intervals of one year, while for Uranus and Neptune the dots show the positions every ten years. As you see, these outer planets require many years to make their circuits about the sun. Indeed, from this diagram you can see

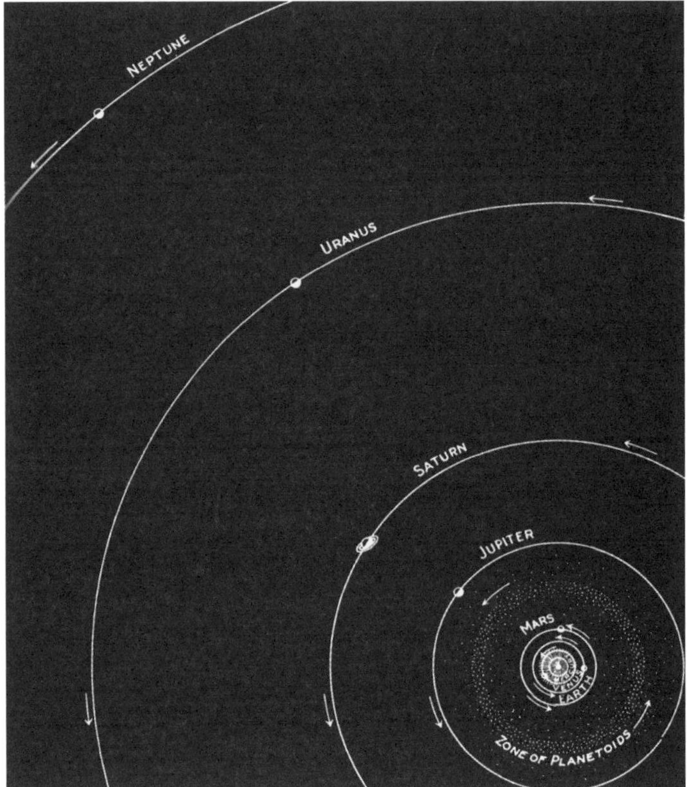

FIG.24 RELATIVE SIZES OF THE ORBITS OF THE PLANETS

The orbits are drawn to scale. Note that Mercury, Venus, the earth, and Mars are close to the sun. Then follows the wide zone of the planetoids or asteroids, and after that are the immense orbits of the giant planets - Jupiter, Saturn, Uranus, and Neptune. They all move in the same direction as their orbits.

how long each planet takes to go completely round its path – Jupiter twelve, Saturn twenty-nine and a half, Uranus eighty-five, Neptune a hundred and sixty-four years.

Each of the orbits shown in Figs. 25 and 26 looks circular in shape, but the real orbits travelled by the planets are ellipses. You know how to describe an ellipse (Fig. 27). You drive two pins in a board, and over them put a loop of string. Put a pencil in the

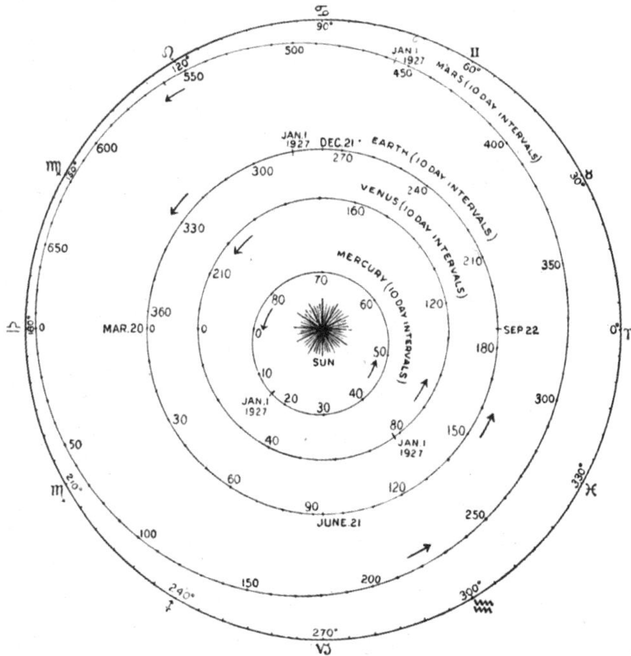

FIG. 25. THE ORBITS OF THE INNER PLANETS
The orbits of Mercury, Venus, the earth, and Mars are here drawn to scale. Their positions are shown on January 1, 1927, and at intervals of ten days.

loop and, keeping the string taut, move the pencil over the paper. The point where each pin is put is called a focus of the ellipse, and if the two pins are near together the ellipse becomes nearly a circle.

As a matter of fact, all the orbits are nearly circular.

There is something else remarkable about the orbits of the planets. They are all very nearly in the same plane (Fig. 28). Further, the planets all travel in the same direction along their orbits. There must surely be some reason for this.

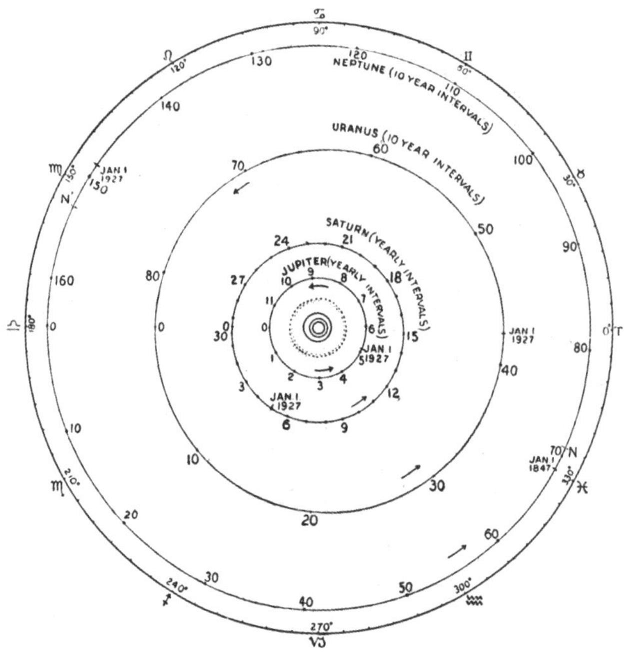

FIG. 26. THE ORBITS OF THE OUTER PLANETS

These orbits are also in proper proportion, but on a different scale from the orbits in Fig. 25, which are shown here in the little circles at the centre. The positions of the planets are given for January 1, 1927, and also at yearly or ten-yearly *intervals.*

In Fig. 29 the relative sizes of the sun and the planets are given. As you see, the four inner planets – Mercury, Venus, the earth, and Mars (shown at the bottom) – are small; while the outer planets – Jupiter, Saturn, Uranus, and Neptune (shown at the top) – are large. Also, the sun is many times as large as all the planets put together.

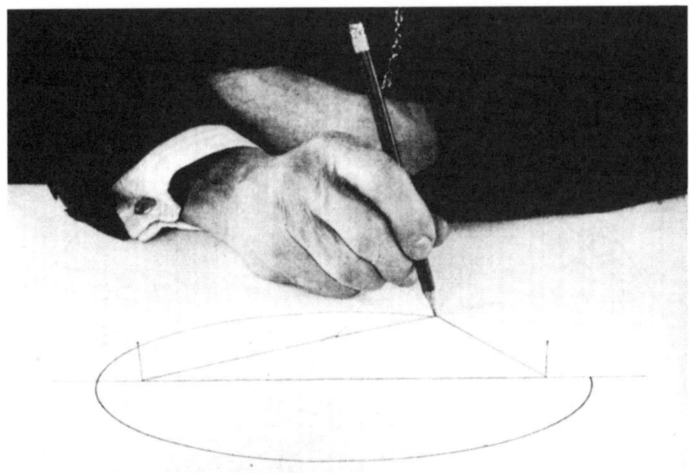

FIG. 27. HOW TO DRAW AN ELLIPSE
If the string is kept taut the pencil will describe an ellipse.
The two pins are at the foci of the ellipse.

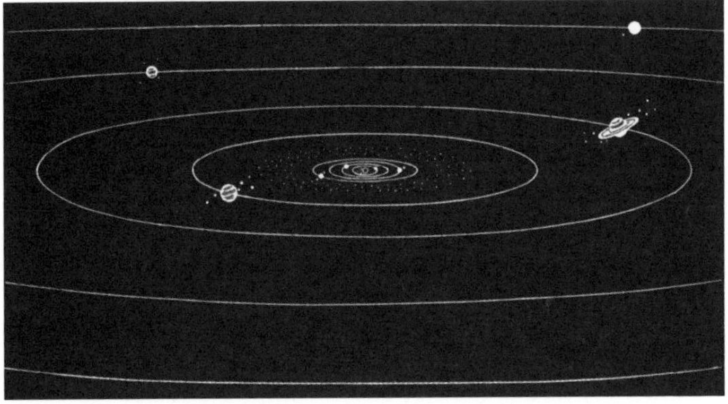

FIG. 28. THE ORBITS OF THE PLANETS IN THE SAME PLANE
This is a perspective view of the planets as they revolve about the sun
in planes which are nearly coincident.

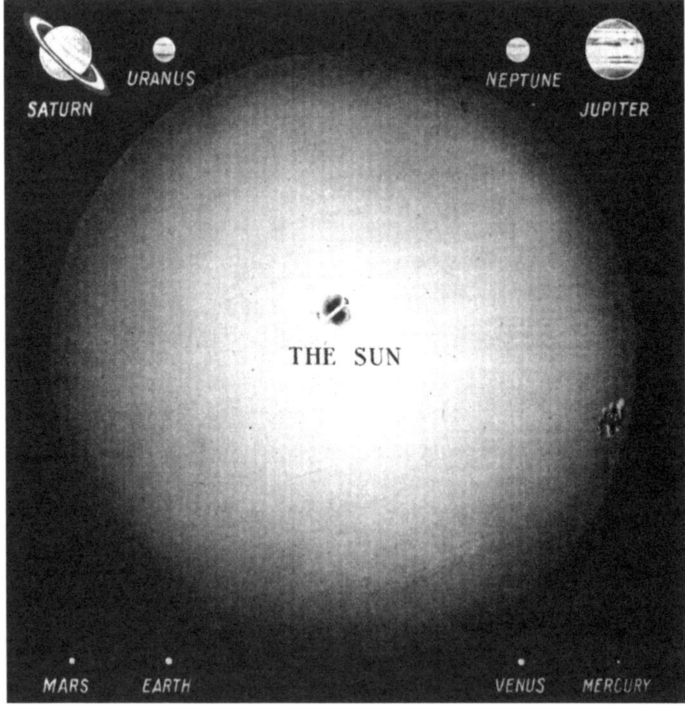

Fig. 29. Relative Sizes of the Sun and the Planets
The inner planets, shown at the bottom, are relatively small. The outer planets, at the top, are much larger. The sun is many times larger than all the planets taken together.
Drawn by F. S. Smith

In Fig. 30 (p. 56) the relative masses, or weights, of these bodies are shown. The great iron weight represents the sun, and the smaller ones below, the planets. As before, the masses of the inner planets are small, those of the outer ones much larger; but the sun is so massive that if we could use up the material in it to make planets 746 sets could be formed out of it.

Having endeavoured to obtain a clear view of the sun's family as a whole, let us next try to obtain better acquainted with the different members. We shall begin with that planet on which we live.

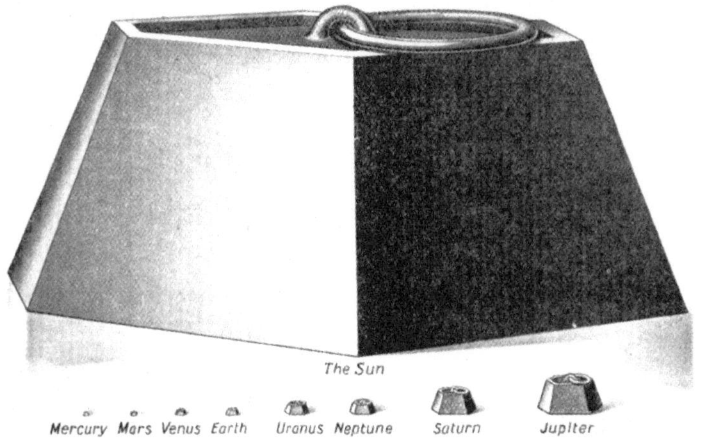

FIG. 30. RELATIVE MASSES OF THE SUN AND THE PLANETS
The great iron weight represents the sun; the little weights, the planets. There is enough matter in the sun to make nearly 750 such sets of planets.

The Earth

The earth is a great ball nearly 8000 miles in diameter. Five hundred years ago it was commonly believed to be flat, and we still occasionally come across people who argue that it is so. There are, however, many observations which lead us to believe that it is spherical in shape.

The first part of a sailing-ship which is nearing her port that a person on shore sees is the tip of the mainmast. In the case of a steamship the smoke from the funnels will first be seen. Next the sails or the funnels come into view, and finally the body of the vessel (Fig. 31). This is precisely what we should expect if the earth were round. If the earth were flat we should see the great body of the ship first.

Again, for £200 or more you may buy a ticket which will carry you completely round the earth. Such cruises are regularly advertised. In the map (Fig. 32) is shown the route which is frequently followed. English travellers would probably leave from Southampton and go southward to Gibraltar. American tourists usually start from New York and cross the Atlantic to Gibraltar, perhaps calling at the Azores islands on the way. Then all pass through the Mediterranean

FIG. 31. SHIPS APPROACHING PORT
As a great ship approaches we first see the tips of the masts, then the sails, and lastly the hull. If the water were a flat plain we should see the hull first.
Drawn by F. S. Smith

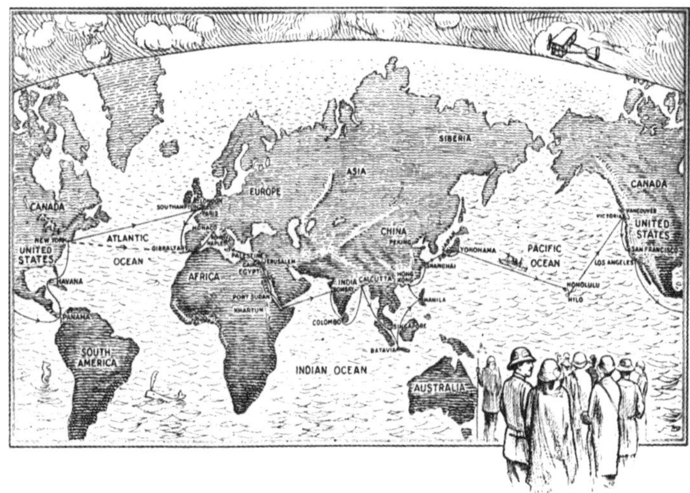

FIG. 32. A CRUISE ROUND THE WORLD
A common route followed by many travellers can be traced on this map. Frequently some of the ports shown here are omitted.

Sea, the Suez Canal, the Red Sea, and the Indian Ocean, to the East Indies and up to Japan, then across the Pacific Ocean to Honolulu, and up to Victoria in British Columbia. After this they go down to San Francisco, then continue along the west coast of North America, pass through the Panama Canal, and go up the east coast to New York. From here the English travellers will cross the Atlantic to their starting-point. To make such a pleasure trip usually requires four months. However, by using railways and aeroplanes, as well as fast steamships, the time can be much reduced. Indeed, the earth has been circled (though not by this route) in a little over twenty-eight and a half days, and undoubtedly this time will soon be shortened.

The Pole Star rises as one goes North

Perhaps some of you have taken a trip to the northern portion of your country and while there have looked for the Pole Star. You found it higher than it appears at your home.

Also some of you may have had the pleasure of going to the Mediterranean Sea or to Florida in the winter time. From there the Pole Star was lower in the sky.

Indeed, we find that for every sixty-nine miles you travel northward, whether on the ocean or on the land, the Pole Star rises 1° in the sky, and if you could reach the North Pole of the earth it would be right overhead.

In the same way for every sixty-nine miles you travel southward the Pole Star sinks 1°, and when you get to the earth's equator it is right down at the horizon and is hardly visible – if you can see it at all.

Now the earth must be practically a sphere, or the Pole Star would not behave in this way. We cannot imagine any other shape the earth could have which would cause the Pole Star to increase and decrease in altitude as it does. Indeed, the astronomer has measured the earth so accurately that he is able to say that it is not an absolutely perfect sphere after all, but that it is slightly flattened at the poles, the diameter from pole to pole being about twenty-seven miles shorter than a diameter at the equator[1].

There are other reasons for believing that the earth is round. The astronomer can compute all the circumstances of an eclipse of the sun thousands of years in advance. He can tell us when the eclipse will begin and when it will end, and what part of the earth you must go to in order to see it. Now he does this on the assumption that the earth is spherical in shape. If it were not his calculations would fail.

Surely then no one who considers the matter can doubt that the earth is spherical in form.

Reasons for believing that the earth is spinning on its axis have already been given (p. 34), and we need not dwell on that subject here.

[1] The equatorial diameter is 7926.68 miles, the polar diameter 7899.98 miles

CHAPTER IV

THE SUN AND THE MOON

The Distance of the Sun from the Earth

EXT let us consider briefly our glorious sun.

There is no other body in the sky which can compare with the sun in its importance to us. We could live without the moon and the stars, but if we were deprived of the sun's light and heat we could not exist many days.

How far away is the sun?

You will naturally wish to know how the distance from the earth to the sun is measured. The method employed is similar to that used by a surveyor, or an explorer, in finding the distance of an inaccessible object.

Suppose you are on the side of a river (Fig. 33[1]), and wish to determine the distance from you to a tree on the opposite bank. You can do this, in an approximate way, if you are supplied with no other measuring instrument than a yard-stick. You might proceed thus:

Obtain a ball of strong string and measure off 50 feet of it. Drive a stake into the ground at A, and, 50 feet away, drive another stake at B. Drive a nail into the top of each stake so that the two nails are exactly 50 feet apart, and stretch the string from one nail to the other. This is our *base line.*

Now while you look from A toward the tree, C, get a friend to drive in a stake at D, and put a nail in the top of it so that A, D, and C are in a straight line. Then, sighting from B toward the tree, have a stake E driven in so that B, E, and C are also in line.

Join the nails in A and D and also those in B and E with stretched strings. If these strings were produced they would meet at C, and, along with the base line, would form a large triangle, CAB.

We know the length of the base, AB, and wish to find the length of the sides AC and BC.

[1] On September 27, 1926, a landslide on the Dent du Midi range, near St Maurice, in Switzerland, blocked the river Rhone for half a mile. The picture shows the river in its new course. After months of labour a canal was dug through the obstruction, and the river was turned into its old channel.

Next, hold a paper under the two strings at A and draw lines on it precisely under the strings. Do the same at B. The lines so drawn will give the angles at A and B – that is, the angles at the base of the triangle, CAB.

FIG. 33. FINDING THE DISTANCE ACROSS A RIVER
By the method shown it is possible to find the distance across a river, the only measuring instrument used being tape-line.
Photograph from "L'Illustration," Paris

Our next task is to draw on a sheet of paper, with great care, a triangle of exactly the same shape as the triangle CAB.

First, draw the line MN (Fig. 34), to correspond to the base line AB, and make it exactly 6 inches long. Then make the angle NMO equal to the angle BAC, and the angle MNO equal to the angle ABC. This must be done very carefully.

The small triangle, OMN, thus obtained is of the same shape as the large triangle, CAB, MO corresponding to AC and NO to BC.

Measure carefully MO and NO. Let us say that they are 19.2 and 18.6 inches respectively.

Now AB, being 50 feet, is a hundred times as long as MN, which is 6 in. Consequently AC must be a hundred times MO – that is, 1920 inches or 160 feet. Similarly BC is 1860 inches or 155 feet.

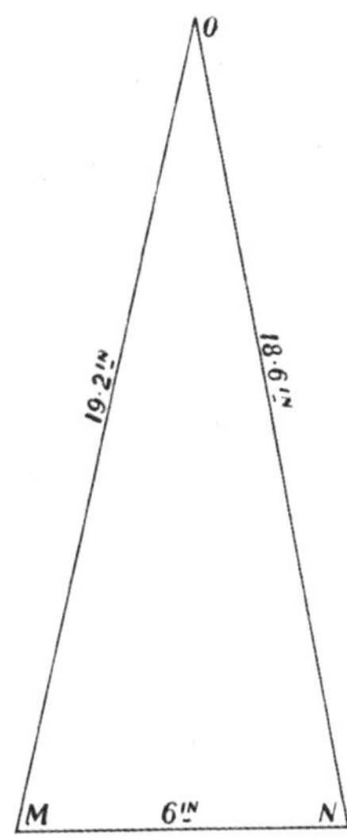

FIG. 34. HOW TO CALCULATE THE DISTANCE ACROSS A RIVER
This triangle is of exactly the same shape as that in Fig.33, and if we know the length of AB we can compute the lengths of the other two sides.

Of course, the surveyor uses instruments by which he can measure lengths and angles exactly. He takes as long a base line as he can, and his results are generally very accurate – his error sometimes being not greater than one inch in ten miles.

Now the astronomer is the man who surveys the celestial distances, but when he tries to apply method to the measurement of the sun's distance he meets very great difficulties, since he has to take his base line on the earth, which is very small compared to the sun's distance. But by using very delicate instruments, exercising extraordinary care, and repeating his observations over and over again, he has at last succeeded in determining how far the sun is from the earth.

It is approximately 93 million miles away.

It is hard to form any idea of this immense distance. If a railway could be built from the earth to the sun, and a train could travel the whole distance at the rate of a mile a minute without stopping during the journey, it would require 175 years to make the journey – two long lifetimes!

The Size of the Sun

How large is the sun?

As soon as we know the distance of the sun it is easy to determine its size. Perhaps you would like to know how it is done. Fig. 35 shows a boy and a girl making an experiment to find out the diameter of the sun. The boy has in his hand a dinner-plate 1 foot in diameter, which he is holding between the girl's eyes and the sun. First, when standing near the girl he holds the plate up, and it covers the sun and quite a bit of the sky as well.

FIG. 35. HOW TO FIND THE SIZE OF THE SUN
The boy is holding up a plate 1 foot in diameter so that it just hides the sun from the girl. Knowing the distance of the boy from the girl and also the distance of the sun, we can compare the diameter of the sun with that of the plate.
Drawn by F. S. Smith

He then moves farther away from the girl, and at last, where he now is, the plate just covers the sun. By means of a long cord the distance from the boy to the girl is found to be 107 feet.

Now let us look at Fig. 36. A represents where the girl's eye is, and XY, at the far side, represents the sun, while BC is the plate, 107 feet

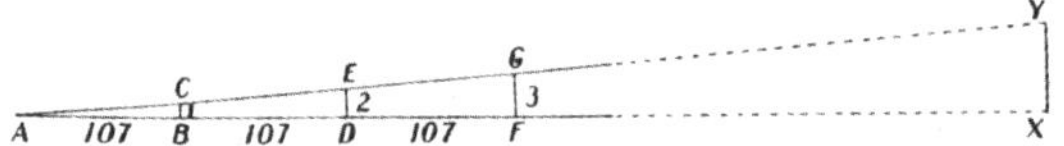

FIG. 36. DIAGRAM SHOWING HOW TO CALCULATE THE SIZE OF THE SUN
Here AB represents the distance of the boy from the girl, and AX the distance of the sun.

away from A, which just hides the sun. It is clear that if the sun was at DE, which is twice 107 feet from A, and was just hidden by the plate, then the diameter of the sun would be twice that of the plate, or just 2 feet. If it was three times 107 feet off – that is, at FG – and was just covered, its diameter would be 3 feet. So you see that the number of times the sun's diameter is greater than that of the plate is just the number of times the sun's distance is greater than 107. Let us divide 107 feet into 93 million miles. It is contained 4589 million times, and so the sun's diameter is 4589 million feet, which is the same as 869,000 miles. More careful measurements make it 866,000 miles. That, then, is the diameter of the sun.

The diameter of the earth is 7918 miles, and by a sum in long division we find that the sun's diameter is nearly 110 times as great as the earth's.

The sun, compared with the earth, is enormous. If the sun were represented by a football the earth would be the size of a small pea!

The width of many of our city streets is 66 feet. Suppose we represent the sun by a great sphere 66 feet in diameter just filling the space where two such roads cross. The earth would be represented by a ball 7 inches in diameter (the size of an ordinary stove-pipe), $1\frac{1}{3}$ miles away.

Sun-spots and Faculæ

If we look at the sun with the naked eye (always, of course through a dark glass) it appears simply as a great bright disc; but with even a small telescope we see interesting features on its surface. Here is a photograph of the sun (Fig. 37). On its disc there are some dark, irregularly shaped markings, which are known as *sun-spots*. Note that the centre part of the spot is darker than the outer portion. Notice also that some of the spots are clustered together in groups, while others are by themselves. Sometimes these spots are so large that several earths could be dropped into them without filling them up.

Just what causes these spots we do not know. You have often heard of volcanoes belching forth fire and smoke and ashes. The earth's crust cracks, and the hot molten material within bursts out. Perhaps something like this happens on the sun. The surface layer becomes weakened at some place; cracks, and allows the fiercely hot matter within to explode and shoot up far above the sun's surface.

In the photograph you will notice that near the edge the disc is darker. This shows that there is an atmosphere of some sort on the sun. You remember, of course, that the sun is a great ball, and what looks to be its 'edge' is that portion of the ball which curves away from us. Now light from that portion passes through more of the sun's atmosphere than does the light from the centre of the disc, and so more of it is lost by absorption. For this reason the edge of the disc appears darker.

Notice also the bright blotches round the group of spots near the edge. These are called *faculæ* (a Latin word meaning "little torches"). They are mountains of flame which thrust their summits above the absorbing atmosphere, just as some mountain peaks on the earth sometimes pierce through a layer of clouds.

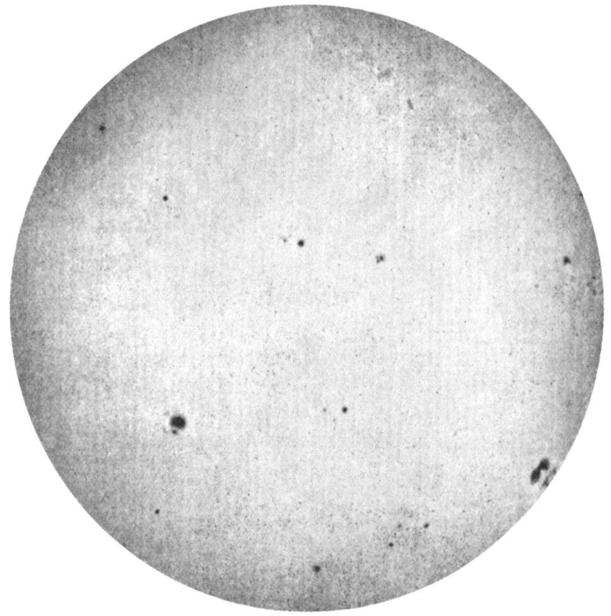

FIG. 37. THE SUN, SHOWING SPOTS AND FACULÆ
Notice the great group of spots, over 125,000 miles long, and the numerous other spots. The fuculæ are the whitish areas seen round those spots which are near the edge of the picture. Note also that the centre of the image is much brighter than the outer portion
Lick Observatory photograph

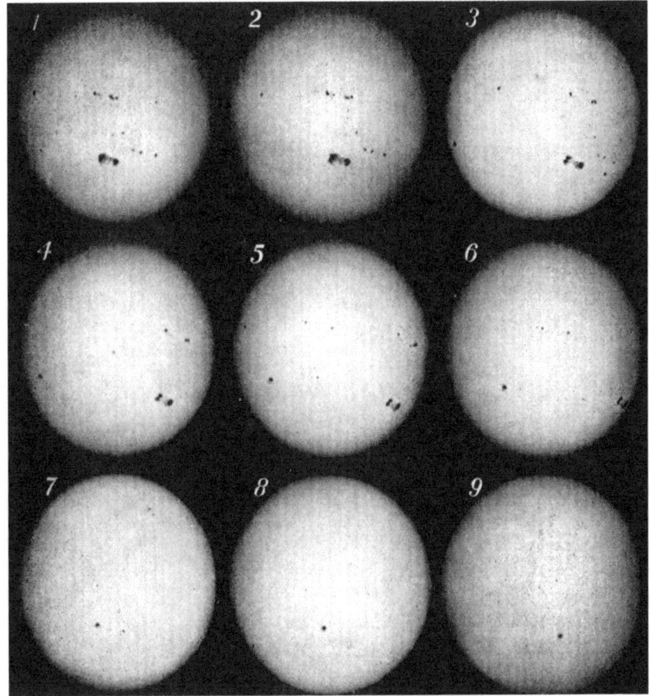

FIG. 38. PHOTOGRAPHS OF THE SUN ON NINE SUCCESSIVE DAYS
By comparing the positions of the spots day after day it becomes evident that the sun is rotating on an axis, and from the motions of the spots the period can be determined. It is about twenty-five days. The photographs were taken on August 6-14 1893, at which time there were many spots on the sun.
Lick Observatory photographs

The Sun rotates on its Axis

In Fig. 38 are nine photographs of the sun, taken on successive days. Look at the large group of sun-spots in the first photograph, which was taken on August 6. We have learned that the diameter of the sun is 866,000 miles. Now this group looks to be about one-tenth as long as the sun's diameter, and hence the group must be about 86,000 miles long. It is so large that it could be seen with the naked eye. Some groups, however, are very much larger.

In the second photograph this group of spots is farther to the right, and if you look for it in the succeeding photographs you see that it moves steadily to the right. In No. 7, taken on August 12, a trace of it can still be seen, but on the next day it has disappeared completely. Other spots behave in just the same way.

Now how do you explain this? You say at once that the sun must rotate on an axis. By observing the spots we can learn where the axis is and how long it takes the sun to turn completely round. It takes about twenty-five days.

The Mass of the Sun

Suppose we were able to take the material of which the sun consists and form earths out of it. How many do you think we could make? There is enough to make 332,000. We say that the mass of the sun is 332,000 times the mass of the earth. Turn back and look at Fig. 30 again.

It is well to know, however, that a cubic foot of the sun would not weigh as much as a cubic foot of earth. Indeed, a cubic foot of water weighs 62½ pounds; a cubic foot of sun 1.4 times as much as a cubic foot of water, or 87½ pounds; while a cubic foot of earth weighs, on the average, 5.5 times as much, or 341 pounds.

Sun-spots, Auroras, Magnetic Storms

Other remarkable facts regarding the sun have been discovered. It has been found that sun-spots are more numerous at some times than at others. In some years – it was the case in 1923 – scarcely a spot can be seen. Then the spots begin to appear more frequently, and in about five years after the time when they are very scarce they become so common that the sun's face is seldom free from these dark blotches. After this they gradually fade away and again almost disappear. This performance of fading away and appearing in full strength again is repeated about nine times in one hundred years.

While these spots are appearing on the sun some strange things are happening on the earth. When there are many spots the magnetic needle suffers many disturbances, and we are said to have *magnetic storms*. At the same time displays of the aurora borealis, or northern lights, are common and brilliant. Now why should these things, which seem to be entirely different in nature, happen at the same time? We do not know; it is still a great mystery.

Photosphere, Chromosphere, Prominences

The bright face of the sun which we see is called the *photosphere*. It is so dazzling that we cannot see anything which may be near it in the sky, just as the brilliant headlights of a car prevent you from seeing anything in their direction. There might be some portion of the sun just outside the photosphere, but our eyes are dazzled and we cannot see it. By means of an instrument called the spectroscope, however, it is possible to explore the edge of the sun's disc, and we find some interesting portions of the sun there.

The layer, or envelope, which overlays the photosphere is called the *chromosphere*, and from it there rise some remarkable shapes called *prominences*. They are usually crimson in colour, and take fantastic shapes. There is much hydrogen gas and calcium vapour in them. Some of these prominences change very little during a week, while

FIG. 39. A PROMINENCE ON THE SUN – THE "HELIOSAURUS"
This strange-looking prominence was visible at the time of the total eclipse of June 8, 1918, the track of which passed over the the United States from Washington State to Florida. This picture was taken at Green River, Wyoming. The earth on the same scale as the prominence is shown in the upper corner.
Photograph by Yerkes Observatory Expedition

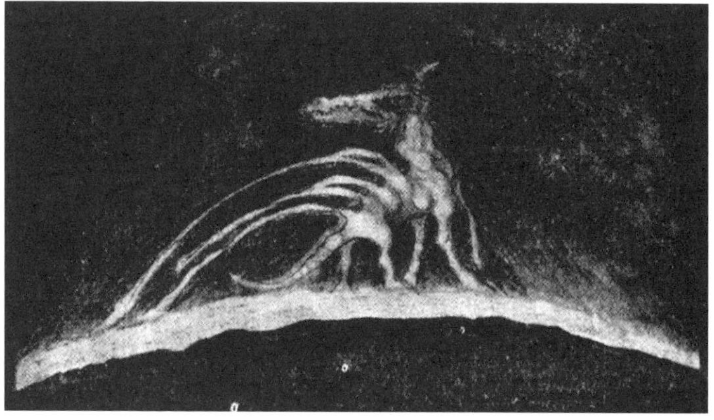

FIG. 40. A DRAWING OF THE "HELIOSAURUS"
The shape of the prominence shown in Fig. 39 suggested this sketch of a fierce solar monster like the dinosaurs of ancient eras. The name means "solar lizard."
Drawn by R. E. De Lury

others show rapid motion. In Fig. 39 is shown a prominence which was photographed on June 8, 1918. It seems to take the shape of a wild monster blowing fire from its nostrils. In order to illustrate this better an astronomer-artist has made a sketch of the fierce-looking animal, with the result shown in Fig. 40. From its peculiar shape this prominence has been named the "Heliosaurus," which means the "solar lizard." The circular spot at the left (Fig. 39) shows the earth on the same scale.

In Fig. 41 are shown seven views of a prominence photographed on October 8, 1920. In the first view, taken at 9.32 T, it was 75,000 miles high; in the last (1.54 P.M.) it was 300,000 miles high; and forty minutes later it reached 517,000 miles, the greatest height ever recorded. In this case also the prominence took fantastic shapes, No. 4 resembling a dog.

These prominences can be seen with the naked eye during a total eclipse of the sun. At such a time the moon comes directly in front of the sun and shuts off the strong light of the photosphere, and we are then able to see the fainter outer portions of the sun.

The Sun's Corona

The way an eclipse of the sun is produced is shown in Fig. 42. Here are our little celestial travellers on a neighbouring planet, watching how the eclipse occurs. The long, slender shadow cast by the moon streams down behind it, striking the earth, and as the moon moves onward in its orbit this shadow trails across the earth's surface. (Think of the moon as coming toward you as you look at the picture.) The strip of the surface over which the shadow moves is called the shadow-path, and if a person is within this path he will not be able to see the sun during the time that the shadow is passing over him – usually for two or three minutes.

One can see at such a time, in addition to the prominences, a wonderful halo of pearly white light surrounding the sun. This is called the sun's *corona* (*corona* being the Latin word for "crown"), and

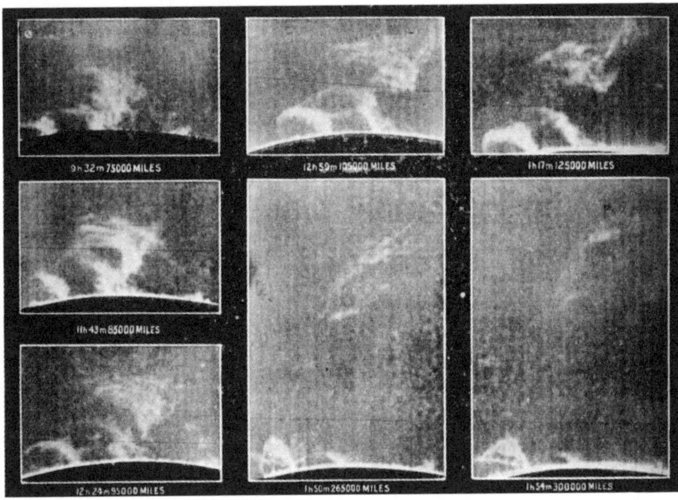

FIG. 41. SEVEN VIEWS OF A GREAT PROMINENCE, OCTOBER 8, 1920
This remarkable prominence assumed various fantastic shapes. one of them roughly resembling a dog. Hence it is known us the "Sun Dog." The earth on the same scale is shown in the upper corner of the first photograph.
Photographs by Pettit, Yerkes Observatory

FIG. 42. HOW AN ECLIPSE OF THE SUN IS PRODUCED
The two little travellers are observing, perhaps from another planet, how an eclipse of the sun is produced. As the moon revolves about the earth (coming out of the picture toward us) its shadow moves along a narrow path on the earth. If a person is within this path the sun will, for him, be entirely covered by the moon during the passage of the shadow over him.
Drawn by Henrietta M. Hopper

it is one of the most beautiful and impressive of celestial spectacles. In Fig. 43 is shown the corona as photographed by the Canadian expedition to the north-west coast of Australia to observe the total eclipse of September 21, 1922. The round, black object you see is the moon; the body of the sun is directly behind it, hidden from our view.

FIG. 43. THE SUN'S CORONA, SEPTEMBER 21, 1922
This photography was taken by the Canadian expedition to eclipse on this date.
The observing station was at Wallal, in South latitude 20°,
on the north-west coast of Australia.

There is something remarkable about the shape of the corona. It changes with the number of spots on the sun's face. When there are many spots the streamers of the corona seem to run out from the sun's surface in all directions; but when the spots are fewer great streamers, or wings, go out from near the sun's equator, and only small ones from its polar regions. These equatorial wings sometimes extend outward two or three times the sun's diameter – *i.e.*, 1½ to 2½ million miles.

What the appearance would be, if we could see the photosphere, chromosphere, and corona of the sun all at the same time, is shown in Figs. 44 and 45. In the first one we are shown the appearance when the spots are very scarce or entirely absent; in the second, when they are numerous. Notice the shape of the corona in the two cases.

Ancient Worship of the Sun

As the sun is the source of our light and heat, and is therefore necessary for the existence of every living thing, it is not surprising that many ancient people actually worshipped it.

Among these were the Egyptians. In their mythology there were as many as two thousand deities, the chief of whom was Ra, the sun-god. Different representations of this god are to be found on

FIG. 44. THE SUN'S PHOTOSPHERE, PROMINENCES, AND CORONA AT A SUN-SPOT MINIMUM
This is a composite picture showing what the sun would look like if its prominences and corona could be seen at the same time as the photosphere. The appearance varies considerably with the number of sun-spots. The view shown in this picture is that when there are very few spots on the sun. Note the great wings of the corona, which extend outward from the sun's equatorial belt a million miles or more. There are some prominences, but not many.
The photographs of the corona and prominences were taken by the Lick Observatory Expedition, located in Georgia, on May 29,1900. A photograph of the photosphere on the same day was supplied by the Royal Observatory, Greenwich.
Drawn by F. S. Smith from photographs

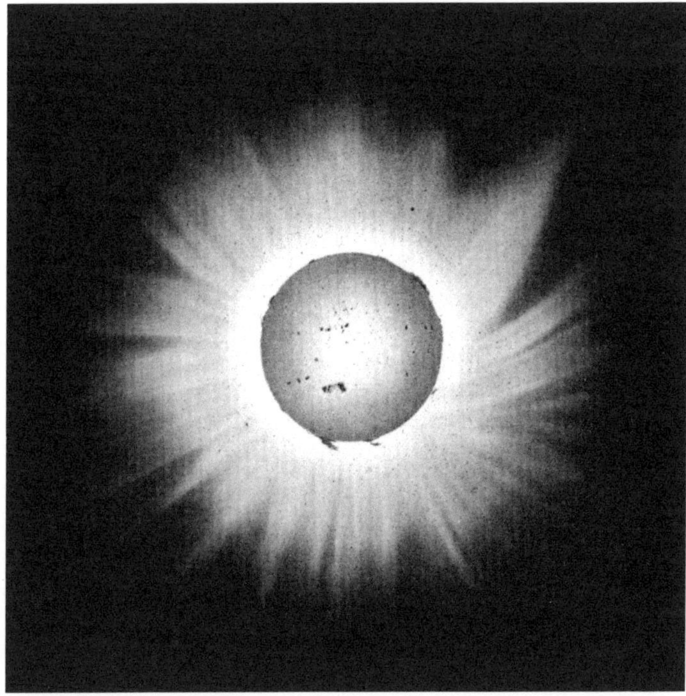

FIG. 45. THE SUN'S PHOTOSPHERE, PROMINENCES, AND CORONA AT A SUN-SPOT MAXIMUM
This is a composite picture similar to the last one, but exhibiting the appearance when the spots are numerous. At that time the streamers of the corona extend outward almost equally from all parts of the sun's surface, and there are many prominences. The photographs of the corona and prominences were taken by the Lick Observatory Expedition to Chile on April 16, 1893. The photosphere is from a Lick Observatory photograph taken on August 9, 1893.
Drawn by F. S. Smith from photographs

the old monuments. One form is shown in Fig. 46. He is pictured with the body of a man and the head of a hawk. There is something impressive in the behaviour of the hawk. At one time it darts downward like a lightning-flash, while at another it soars gracefully on outstretched wings in the high heavens. Quite naturally it was associated with the sun, and was looked upon as sacred. Upon his head Ra bears the solar disc, about which is the *urœus*, or asp – a sign of royalty in old Egypt. In his right hand is the *ankh*, which is the sign of life, and in his left is the sceptre, the symbol of power. Above the figure is the name of the god written in hieroglyphics, or picture-language. A circle with a dot within it ⊙ is still used by astronomers as a symbol for the sun.

FIG. 46. RA, THE EGYPTIAN SUN-GOD
Representations of Ra, the great sun-god, are found on many ancient Egyptian monuments and temples.

What is the Sun made of?

Though the sun is so far away astronomers have found out what it is composed of. This has been done by means of the spectroscope, the wonderful instrument which has already been mentioned (p. 68).

And what do you think the sun is made of? It is composed of iron, copper, zinc, sodium, calcium, hydrogen, and many other substances which we find here on the earth.

Is not that remarkable? The sun and the earth made of the very same materials! Surely they were joined together in a single mass at a time in the far-distant past.

The Moon – its Distance and Size

When we compare the sun and the moon up in the sky they look to be of about the same size, and you may be surprised to learn that the sun is actually 400 times as large as the moon. It appears of about the same size because it is 400 times as far away.

The moon's distance from the earth is, on the average, 239,000 miles. This is small compared to most of the distances which we meet with in astronomy. An engine-driver taking his train from London to

Berwick or Toronto to Montreal (330 miles) six times a week would travel as far as the moon in two years, but to go as far as the sun would require 800 years.

The moon's diameter is 2160 miles, a little more than a quarter that of the earth, and there is enough material in the earth to make 81 moons. Their relative sizes are shown in Fig. 47.

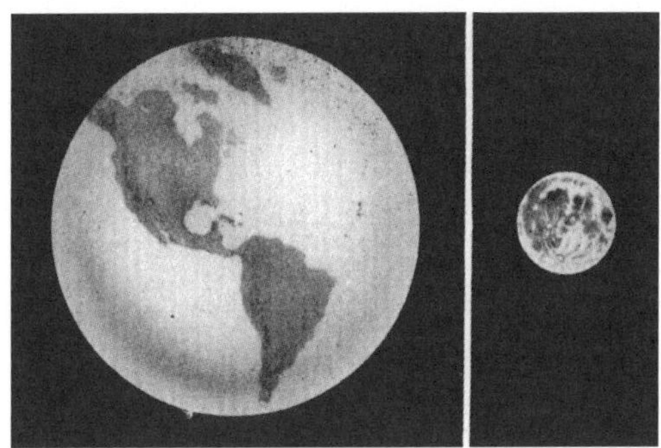

FIG. 47. RELATIVE SIZES OF THE EARTH AND THE MOON
The diameters are 7918 miles and 2160 miles – *i.e.*, they are in the ratio of 100 : 27.

The Phases of the Moon

As we have already learned (p. 36), the moon revolves about the earth. You must remember also that the moon is a dark body and can be seen only as it is lighted up by the sun. It is on account of these two facts that the moon shows phases. You must understand how they are produced, and the explanation is easy.

In looking at Fig. 48 you must think of the sun as being far off to the right. You see the rays coming from it. When they fall upon the moon they illuminate one-half of it, just as they illuminate one-half of the earth or any other round body on which they fall.

The moon revolves about the earth in the direction indicated by the arrows. When it is at A in the line drawn from the earth to the sun the bright face of the moon is turned from the earth, and we cannot see it at all. When in this position the moon is said to be *new*.

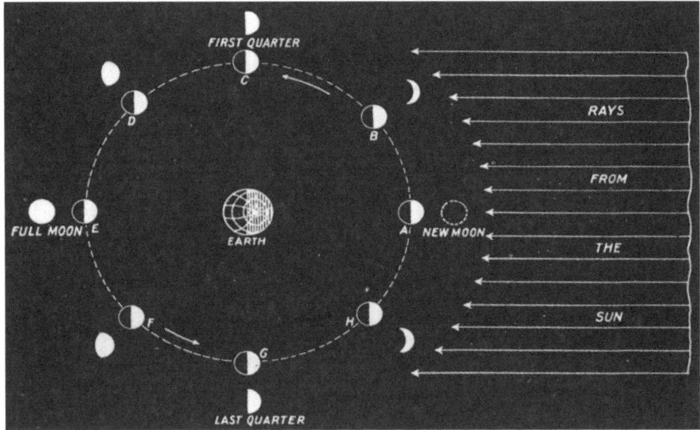

FIG. 48. HOW THE PHASES OF THE MOON ARE PRODUCED
The sun is far away to the right. Its rays at all times illuminate one-half of the moon's surface, and of this we see varying portions us it revolves about the earth. At A we cannot see any of the bright hemisphere, at E we see it all, at C and G we see half of it.

About three days later the moon reaches position B. A person on the earth can now see a portion of the illuminated hemisphere of the moon, but not the whole of it. The part he sees looks like a crescent, as shown in the diagram. The moon is said to be three days old.

About four days later the moon arrives at C, and from the earth we see one-half of the illuminated face. It looks like a bright half-circular disc. The moon has now travelled one quarter of its orbit, counting from when it was new, and it is said to be at its *first quarter*.

By the time the moon is ten days old it has reached D, and its shape is said to be *gibbous*, which means "swelling out." About two weeks after the moon is new it is found at E. From the earth we see its entire bright hemisphere, and it is said to be *full*. When it looks like this the moon, the earth, and the sun are in line, the earth being between the other two bodies. As the sun sets in the west the full moon rises in the east, and at midnight it is on the meridian.

A week later the moon reaches G. Its shape is now semicircular again, and it is at its *third* or *last quarter*. In about one week's time, or after a month in all, it arrives at A once more, and we have *new moon* again. The complete performance, from one new moon to the next, requires twenty-nine and a half days.

When the moon is strictly new we cannot see it at all. It has to be about two days old before we can see its slender crescent. About two days before it becomes new, or when it is about twenty-seven days old, it shows a similar crescent. At this time one must look for it in the east, just before the sun comes up. When it is young – two or

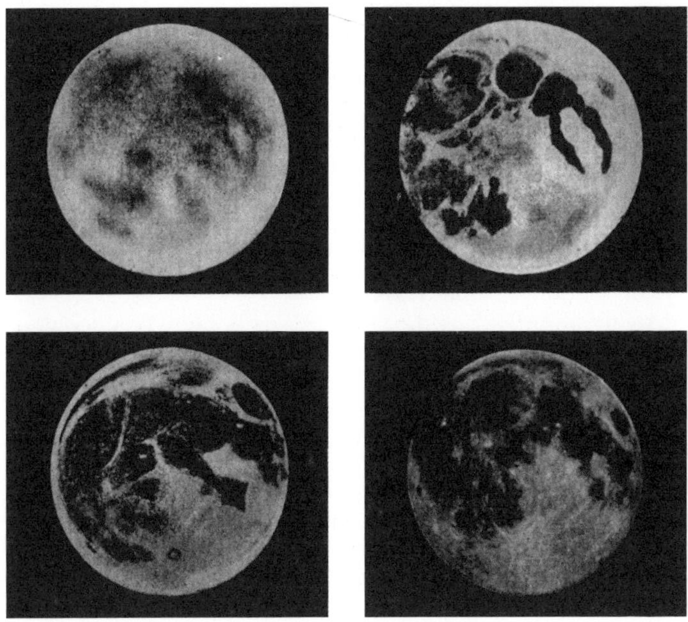

Fig. 49. Naked-eye Views of the Full Moon

In the upper left-hand picture is a drawing of the 'man in the moon'; in the upper right, the 'crab'; in the lower left, the 'woman reading a book.' In the lower right-hand picture is an actual photograph of the moon, with which the other pictures may be compared. A lively imagination will recognize other figures in the moon.

By permission from "The Moon," by W. H. Pickering

three days old – we must look for it in the west just after the sun has gone down.

The Moon and the Weather

Remember that the horns of the crescent moon are always turned away from the sun. Fig. 48 shows that this must be so. Further, the line joining the tips of the horns is at right angles to the line joining the moon and the sun. In this part of the earth where we live, which is known as the middle latitudes, the horns in the springtime are turned upward so that the moon looks as though it would hold water; in the autumn they are tilted so that the water would be spilt.

In the tropical parts of the earth the horns of the crescent moon are always turned so that it could hold water, while in the arctic regions they are tilted so that water would run out. In neither position has the new moon anything to do with wet weather. You see how ridiculous it is to connect the weather with the appearance of the Crescent moon. Indeed, though many people believe that there is a connection, the moon has absolutely no effect on the weather at all.

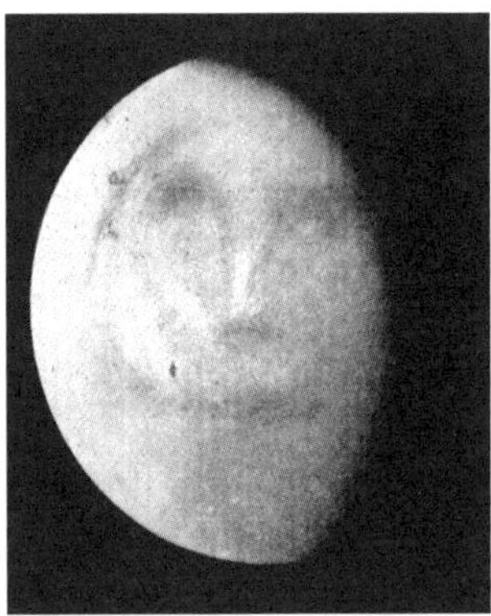

FIG. 50. THE 'MAN IN THE MOON'
The 'man' can also be seen easily when the moon is about half-way between full and third quarter and when it is not far above the eastern horizon. This picture is from a sketch made on September 14, 1927, when the moon's age was nineteen days.

Naked-eye Views of the Moon

It is interesting to study the moon with the naked eye. In Fig. 49 there are four pictures. On the lower right-hand side is a photograph of the full moon which illustrates well how the moon looks when viewed with a small telescope or a field-glass. In the upper left-hand picture is seen the 'man in the moon.' Notice his eyes, nose, mouth, and chin. As the full moon is coming up in the autumn the 'man in the moon' is well seen. He is also clearly seen when the moon is about four days past full (Fig. 50). At the lower left (Fig. 49) is shown the 'woman reading.' She has a hat on, and is bending over and holding the book up before her. Then in the upper right picture is seen the 'crab.' Other fanciful objects have been sketched by different observers. Those shown here are very easy to see – if you look for them.

FIG. 51. THE FULL MOON

The dark areas are called 'seas,' though they contain no water; the small, round objects are craters, probably volcanic in origin. Note the bright 'rays,' many miles long, coming from the crater Tycho at the upper part of the picture. No one has satisfactorily explained these.

Harvard Observatory photograph

FIG. 52. THE MOON AT FIRST QUARTER
The dark seas are well shown here, and the craters are very prominent along the ragged edge (known as the "terminator") which bounds the visible portion. The largest one, half-way down this edge, is about 115 miles in diameter – fifteen times as large as any crator on the earth. The other half of the moon's far is in darkness, and so cannot be seen by us. *Yerkes Observatory photograph.*

FIG. 53. THE SEA OF SHOWERS AND THE SURROUNDING SURFACE
This photograph was taken at Mount Wilson, California, with the Hooker telescope, which is 100 inches in diameter – the largest in existence. Observe the Sea of Showers, the mountain ranges, and the craters; also the sharp shadows cast by the rims of the craters and by rocky objects on the great sea-bottom.
Mount Wilson Observatory photograph

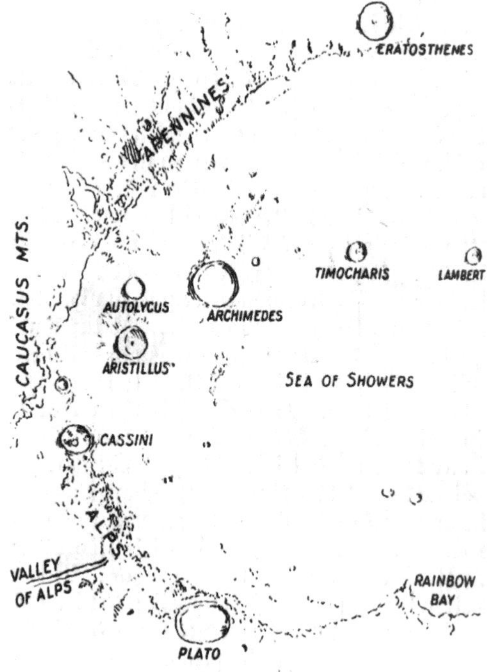

FIG. 54. KEY TO THE FEATURES SHOWN IN FIG. 53
The craters are named mostly after noted astronomers, and the mountains after mountains on the earth.

Photographs of the Moon

In Fig. 51 we have a photograph of the full moon. The illustration is turned to show the moon as it is seen in a telescope – that is, upside down as compared to the view with the naked eye or with a field-glass.

Note the dark portions, mostly roundish in shape. They are called *seas*. When first discovered they were supposed to be bodies of water, and, although we now know that there is no water on the moon, they are still called seas. They bear fanciful names. The oval spot half-way up on the left side is the Sea of Crises. Immediately to the right of it is the Sea of Tranquillity, and directly below this, also oval in shape, is the Sea of Serenity. The big one below and to the right is the Sea of Showers. Notice also the bright spot near the top, with bright rays running from it. It is not a sea, but a crater, and it is called Tycho, after a great Danish astronomer.

Next we have a photograph (Fig. 52) of the moon at its first quarter, or when it is about seven days old. Some of the seas are shown very well, but notice especially the numerous round objects, seen particularly clearly along the right-hand edge. These are craters, perhaps due to volcanic action in long-past ages. The largest one on the moon is seen half-way down the right-hand edge. It is called Ptolemy, after another great astronomer, and is 115 miles across. The largest crater on the earth, which is found in Japan, is only seven miles in diameter. Look also at some detached white spots in the darkness just beyond the ragged edge. These are the tops of mountain-peaks which are just being lighted up by the rising sun.

Fig. 53 is from a photograph taken with the largest telescope in the world, which is on Mount Wilson in Southern California. This shows a portion of the moon (the northerly part) as seen at third quarter. Here is the Sea of Showers. The names of some of the objects seen in this picture are given in the key map (Fig. 54). First we have the Sea of Showers. On its upper left shore are the Apennines, a range of mountains named after the Apennine Mountains in Italy. At the lower left are the Alps, while the Caucasus Mountains are between. Notice the valley in the Alps. It is from three to six miles wide and over eighty miles long. Then there is the crater Plato. It looks elliptical in shape, because it is on the surface which is curving away from us, but in reality it is circular. It is sixty miles in diameter. Then there is Rainbow Bay, and higher up are the craters Archimedes, Eratosthenes, and others. Also note the sharp rock rising up at the left side of the Sea of Showers. It throws a very black shadow. This shows in what direction the sun is. It is interesting to pick out all these objects on the photograph. There are many other objects, and a name has been given to all the principal ones.

Does the Moon always show the Same Face

Month after month as we look at the full moon the same familiar features are seen, and so we are in the habit of saying that the moon always shows the same face to the earth. And, indeed, that is practically correct. Should Hipparchus, the distinguished Greek astronomer, who lived more than two thousand years ago, visit the earth now, he would see the same face of the moon which he gazed upon from his observatory on the island of Rhodes in the Ægean Sea.

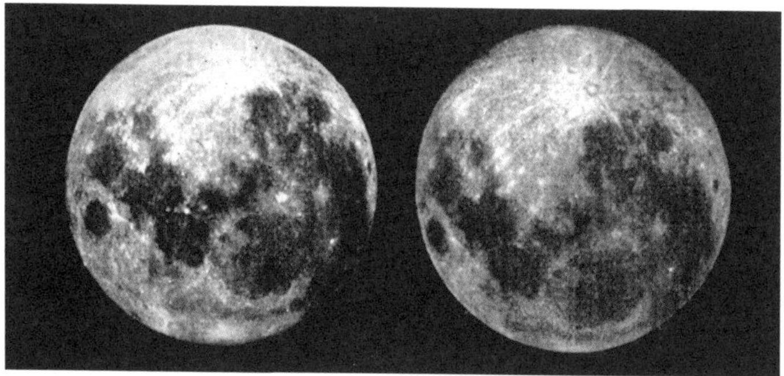

FIG. 55. TWO PHOTOGRAPHS OF THE MOON, SHOWING LIBRATION
A comparison of these two photographs, the second of which was taken after an interval of six months (October 31, 1906, and April 27, 1907), shows that the face of the moon which we see is not always exactly the same. The bright spot near the top is the crater Tycho, and the bright radiations from it are conspicuous.
Amateur photographs, D. B. Marsh

When, however, we examine it closely we find that the portion of the moon's surface which we see is not always exactly the same. This is well illustrated in the two photographs of the full moon in Fig. 55. Compare them carefully. In the right-hand view we see much farther above the crater Tycho, and, of course, much less below the Sea of Showers, than in the left-hand one. Also notice that a portion to the left of the Sea of Crises is visible in the left-hand picture which cannot be seen in the other one.

Indeed, by continual observation with the telescope, or by studying photographs taken at different times, it is possible to see 59 per cent of the entire surface of the moon. Some people express a wish to see the 41 per cent. which is always hidden from view, but they may rest assured that it is of the same general nature as that which we are permitted to see.

Worship of the Moon

The moon also was worshipped by the Egyptians. They had two moon-gods, one of them bearing the name of Thoth. One form in which he was pictured is shown in Fig. 56. He has the body of a man and the head of an ibis – a bird regarded as sacred in ancient Egypt. Upon the head is a disc and crescent, to which an ostrich feather, the symbol of truth, is added. In his hands he carries a writing-tablet and a reed pen. Thoth was also the god of science and letters.

The Moon a Dead World

It is the opinion of astronomers that the moon is a waste of rock and sand. No clouds are ever seen upon its surface, and no changes of its features have been surely detected. They look somewhat different according as they are lighted up differently by the sun's rays, but there is no real change in their shapes. Hence we believe that there is no atmosphere on the moon. When the sun's radiation beats down upon its surface during its daytime it must become very hot, but during its long night the temperature must fall very low.

All is dead, deserted, and silent there.

FIG. 56. THOTH, THE EGYPTIAN MOON-GOD
This is one of the forms of Thoth found on ancient monuments. Thoth was also the god of science and literature.

CHAPTER V

MERCURY AND VENUS

Mercury

E now come to the planets. The one nearest to the sun is called Mercury. Its average distance from that body is 36 million miles, though actually it varies between 28½ and 43½ millions. As we have already learned, it is the smallest of the family.

In the ancient Greek and Roman mythology Mercury occupied the position of messenger of the gods. He was represented as a fine youth with wings on his heels which enabled him to rush through space with tremendous speed when he was performing his official duties. He was a favourite subject for artists and sculptors, and many fine paintings and statues of him have been made. A photograph of one of the most graceful and charming of these statues is shown in Fig. 57. It was made by Giovanni da Bologna in Italy about 1575, and is in an art gallery in Florence.

FIG. 57. MERCURY, THE MESSENGER OF THE GODS
In his left hand he bears his official wand, or sceptre. His left foot is supported by the west wind.

The planets are named after the old heathen gods, and the name of Mercury was naturally given to the nearest planet, since it moves the swiftest of all. It never travels at less than twenty-three miles a second, and it sometimes reaches thirty-five miles a second. It requires only eighty-eight days to make its circuit about the sun, or, in other words, Mercury's year is not quite three of our months. Its diameter is 3030 miles, and its mass is about one-twentieth that of the earth. The relative sizes of the earth and Mercury are shown in Fig. 58.

Astronomers for a long time have patiently watched the planet with the telescope in the hope of recognizing some marks on its surface which would allow them to determine how long it takes Mercury to rotate on its axis, but they have not been very successful. It is generally believed that Mercury always presents

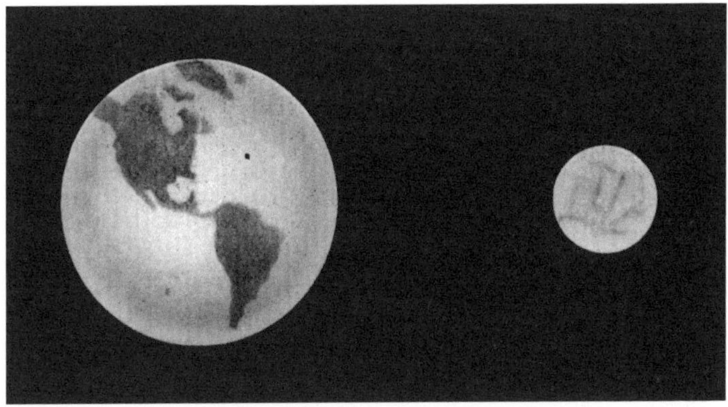

FIG. 58. RELATIVE SIZES OF THE EARTH AND MERCURY
The diameters are 7918 miles and 3030 miles – *i.e.*, they are in the ratio of 100:39.

the same face to the sun, and that it has little or no atmosphere. Therefore, on one face it must be a burning desert, on the other a frozen waste.

Venus

The planet next in order is Venus. Its distance from the sun is 67 million miles. Just as Venus was the most beautiful of all the ancient deities, so Venus is the loveliest of all the sun's family. As an evening star in the west, or a morning star in the east, Venus easily outshines all the other planets and all the stars, and draws forth the admiration of every one.

At the speed of twenty-two miles per second Venus pursues her path, and makes a complete revolution in 225 days or seven and a half months. The diameter of the planet is 7700 miles, which is almost the same as that of the earth. Indeed, Venus and the earth are twin sisters (Fig. 59).

Before we consider the behaviour of Venus there are some things which should be remembered:

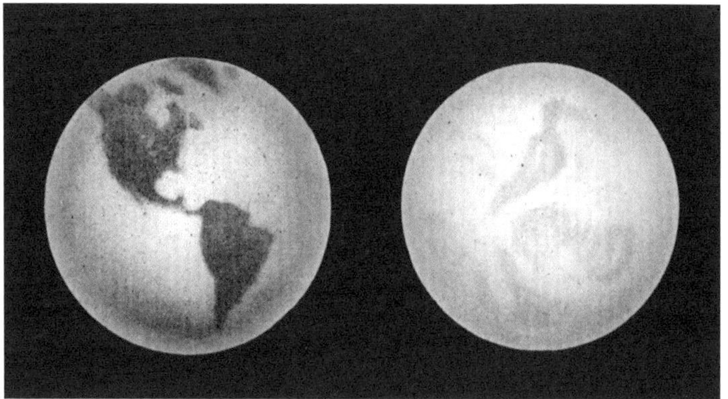

FIG. 59. RELATIVE SIZES OF THE EARTH AND VENUS
The diameters are 7918 and 7700 miles – *i.e.*, they are in the ratio of 100:97.

(1) Venus has no light of its own, but shines only as it is illuminated by the sun.

(2) At all times one-half of the planet's surface is illuminated – namely, that hemisphere which is turned toward the sun.

(3) We observe it from the earth, which is 93 million miles from the sun, and therefore beyond the orbit of Venus.

In Fig. 60 we see a young astronomer on the earth watching Venus as it moves round the sun. When Venus is at position 1 it is at its greatest distance from the earth, being then 93 plus 67, or 160, million miles away. This is a very great distance indeed, and consequently the planet appears small. The planet really cannot be observed when it is in this position, as it is lost in the blaze of the sun's light. As observed from the earth, it will usually be a little north or a little south of the sun, but so close to it in the sky that it cannot be seen. If at this time the moon comes in front of the sun and shuts off its intense light the planet can be seen, and through a telescope it shows a bright round disc. This is, of course, exactly what we should expect, since its illuminated hemisphere is turned directly toward us. Indeed, it has been photographed when very close to this position, and it shows a round disc (Fig. 61, A).

When it reaches 2 (Fig. 60) it is nearer the earth and for that reason looks larger, but now we cannot see the entire illuminated hemisphere. With the naked eye one cannot at any time make out the shape of the disc, but a small telescope will show that it has the gibbous form shown in Fig. 60. On reaching 3 it is much closer and brighter, and it looks like a half-moon.

As it moves along its orbit it becomes still brighter until it comes to 4. The telescope reveals it now to be crescent-shaped. (Compare Fig. 61, B, C, D.)

Five weeks later the planet reaches 5 (Fig. 60). At this time its distance from the earth is 93 minus 67, or 26, million miles. It is

FIG. 47. RELATIVE SIZES OF THE EARTH AND THE MOON
The young astronomer on the earth watches Venus as it moves from superior conjunction with the sun (position 1) to inferior conjunction (position 5). It appears to increase in size mid to show phases.
Drawn by F. S. Smith

FIG. 61. PHOTOGRAPHS OF VENUS
Here we have actual photographs of the planet, showing the increase in apparent diameter and also the phases, which vary from 'full' to it slender crescent.
Photographs by E. C. Slither, Lowell Observatory

now closest to the earth, but, as its illuminated face is turned from the earth, we cannot see the planet at all. Venus hides her beautiful face from us.

Then in five weeks more the planet moves to 6, where its shape is similar to that at 4. Then it moves to 7 and 8 and 1 again.

We thus see that Venus shows phases like the moon.

Morning and Evening Star

We have not finished with this picture yet. When Venus is at 2, 3, or 4 a person on the earth will see the planet up in the sky to the left-hand of the sun. Look up now at the sun and think of Venus as being on your left-hand – that is, east of the sun. Then as the sun moves across the sky during the day Venus will follow it, and when the sun has set in the west Venus will still be above the horizon, and we call it an *evening star.*

When, however, Venus is at 6, 7, or 8 it will appear in the sky to the right, or west, of the sun. Hence as the sun moves across the sky the planet will move along ahead of it (will precede it, as the astronomer says), and so will set before the sun; but next morning it will come up in the east before the sun – that is, it will be a *morning star.*

The ancient people thought the evening and the morning star were two different bodies, and called them Hesperus and Phosphorus respectively.

FIG. 62. DR PERCIVAL LOWELL OBSERVING VENUS IN THE DAYTIME
The best time to observe Venus is in the daytime, when it is high above the horizon. Dr Lowell is using the 2-foot telescope of the Lowell Observatory, Flagstaff, Arizona, which he founded. It was with this telescope that Dr Lowell made most of his observations on Mars and other planets.

No definite permanent markings on the surface of Venus can be seen, and it is therefore difficult to determine how long it takes to turn on its axis. For a long time it was believed that the planet rotated in a little less than twenty-four hours. Then further observations led to the view that it presented the same face always to the sun, and therefore that its rotation and revolution periods were the same – namely, 225 days. If the rotation period were in the neighbourhood of a day the planet would be flattened at the poles. This effect is shown by all the planets which are known to have short periods. But many accurate measurements have not revealed any difference between the polar and equatorial diameters of Venus. At present the evidence is favourable to a period longer than twenty days – perhaps five or six weeks.

Venus has a thick atmosphere which prevents the surface from getting either too hot or too cold. Mercury has no such protection. It is this atmosphere which prevents us from seeing any marks on the solid surface of the planet.

Venus seen in Daylight

Venus when at its greatest brilliancy far outshines any other planet or star, and can be seen easily in full sunlight if one knows just in what direction to look for it.

Comparatively few people have seen the planet in the daytime, but on some occasions it has aroused much interest. In 1716 many citizens of London were thrown into a state of excitement on seeing a strange object in the sky in the daytime, and feared that it might be the forerunner of some disaster. But the astronomer Halley explained that it was simply the planet Venus following its usual course.

It is also reported that on one occasion, when Napoleon was on his way to a state ceremony in Paris at noon, he was surprised to find the people more interested in looking at something in the sky than in himself and his brilliant staff. On seeking the cause the planet was pointed out to him.

Quite recently there was some stir in the city of Los Angeles, in California, when the people gathered in considerable numbers to see Venus. The tall buildings kept out the direct sunshine and permitted a comfortable view of that part of the sky in which the planet was. But it may be seen quite well in the country. A few years ago some persons living on the prairie in Western Canada, on discovering the planet in the sky, telegraphed the news throughout the country.

Venus is always a dazzling object in the telescope, and is best observed when high above the horizon. In Fig. 62 Dr Percival Lowell, of the Lowell Observatory, Flagstaff, Arizona, is shown observing Venus in the daytime. When brightest its shape is like the moon five days old – like that between 3 and 4 or 6 and 7 in Fig. 60.

CHAPTER VI

MARS

The Planet's Distance and Size

F all the planets Mars has excited the greatest interest. It seems to be more like the earth than any of the others, and everybody is anxious to learn if there are living beings upon it.

The planet's distance from the sun is about 142 million miles, which is one and a half times the earth's distance, and it completes its circuit about the sun in 687 days, travelling continually with a speed of fifteen miles a second.

The diameter of Mars is 4215 miles, not much more than half that of the earth (Fig. 63), and its mass is one-ninth that of the earth. In other words, there is enough material in the earth to make nine planets like Mars.

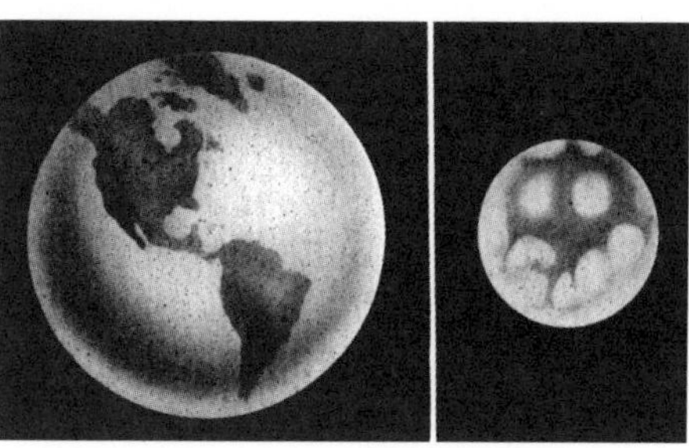

FIG. 63. RELATIVE SIZES OF THE EARTH AND MARS
The diameters are 7918 miles and 4215 miles – *i.e.*, they are in the ratio of 100 : 53.

Orbits of the Earth and Mars

Let us look into the motion of the earth and of Mars. In Fig. 64 the inner orbit is that of the earth, the outer is that of Mars. The orbits are really ellipses, but they do not differ much from circles.

The position of the earth on the first day of each month is given. It occupies these same places year after year.

The position of Mars on the first day of each month of 1926 is also shown. As it takes almost two years to go round the sun, it passes over only a little more than half of its orbit in one year. The line joining the two bodies represents the distance between them on the different dates, and it is interesting to see how the distance changes. The figures beside the line give the distance in millions of miles.

On January 1 the planet was 208 million miles from the earth, but by February 1 this had been reduced to 188 million. Thus during January Mars came closer to the earth by 20 million miles, or the speed of approach was about ten miles per second. The distance between the bodies continually diminishes until

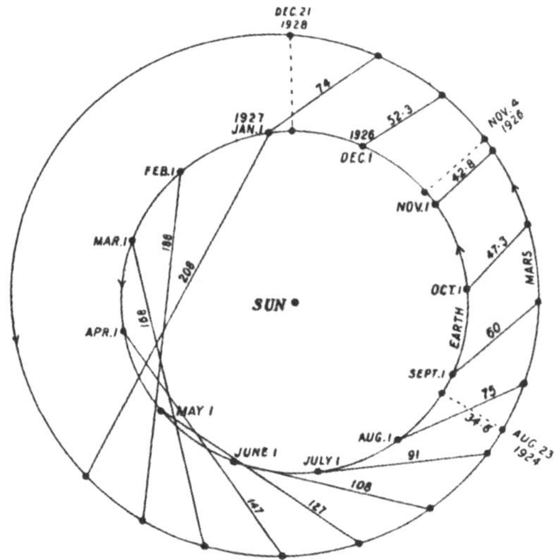

FIG. 64. DIAGRAM OF THE ORBITS OF THE EARTH AND MARS
The orbits are drawn to scale. Note that the orbits are closer together at one side than at the other. The position of the earth on the first of each month every year, and of Mars on the first of each month of 1926, is shown, and the number beside the line joining the two bodies expresses the distance apart in millions of miles.

November. On the 4th of that month Mars and the earth are in the positions shown. At this time the sun, the earth, and Mars are in line, and a person on the earth sees the sun and the planet on sides of him. The astronomer describes this condition of things by saying that Mars is in opposition to the sun. As the sun sets in the west Mars rises in the east, and it is a prominent object in the sky all night long. At the opposition of Mars of November 4, 1926, Mars was about 43 million miles from the earth.

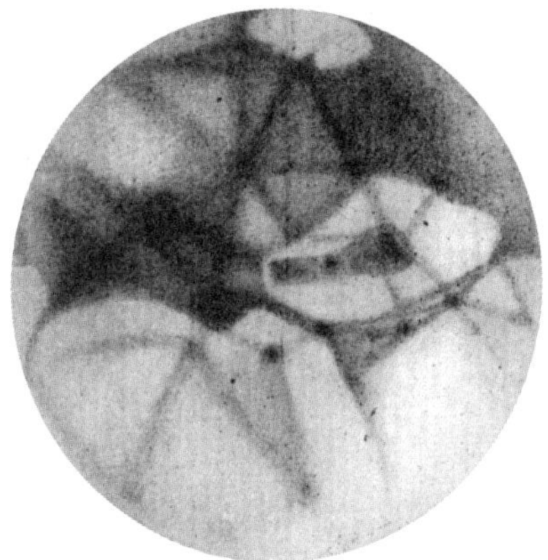

FIG. 65. A DRAWING OF MARS

This drawing represents Mars as seen with the 36-inch telescope of the Lick Observatory, in California, on August 29, 1924. The bright spot at the top is the pole cap, the darker portion (greenish in the telescope) is probably vegetation, and the lighter portion (pinkish in the telescope) desert. The straight streaks are "canals."

Drawn by R. J. Trumpler

The interval from one opposition to the next is about 780 days. Thus there was an opposition on August 23, 1924, at which time the distance from Mars to the earth was less than 35 million miles. The diagram shows where they were then. This was the closest approach of the two bodies for many years past or to come. The opposition next after that of November 4, 1926, is about 780 days after that date – about December 21, 1928, and the distance then will be about 55 million miles.

The Appearance of Mars in a Telescope

When Mars is at a very great distance from the earth, as it was in January 1926, it is not very bright – about as bright as the Pole Star; but when it is near it becomes one of the most brilliant objects in the sky. It has a ruddy, fiery colour, and one is not surprised that the ancients named it after their god of war.

When a large telescope is turned on Mars we see interesting features on its surface. It is quite unlike Mercury or Venus, on which we cannot see any definite markings at all. Fig. 65 shows a drawing made at the Lick Observatory in California in 1924. Certain details on the surface are clearly shown. When viewed a few hours later the face is quite different. This fact is well illustrated in Fig. 66. The right-hand photograph in this was taken at 9.24 P.M., the middle one at 10.24, and the left-hand one at 10.46. There is a decided change between the first and the last, due to the rotation of the planet during the hour and twenty-two minutes which elapsed between the exposures.

By watching the markings on the planet the time required for a complete rotation has been found. It is 24 hr. 37 min. 22.6 sec. This is the length of a day on Mars.

The planet also has seasons, but they are nearly twice as long as ours. They are also much colder, since Mars is one and a half times as far from the sun, which is the source of all the heat received both by the earth and Mars. In the long winter the temperature falls very low, and beings such as ourselves could hardly exist there.

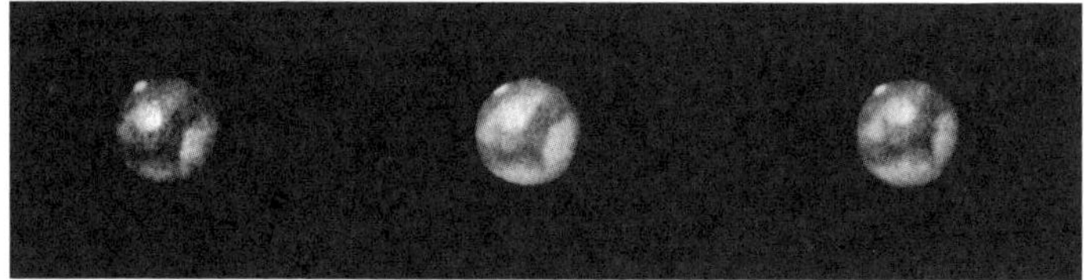

FIG. 66. MARS: THREE PHOTOGRAPHS SHOWING ROTATION
These photographs, from right to left, were taken at 9.24, 10.24, and 10.46 P.M. respectively. In the interval of an hour and twenty-two minutes between the first and the third photograph the rotation of the planet is clearly seen.
By Barnard, Yerkes Observatory

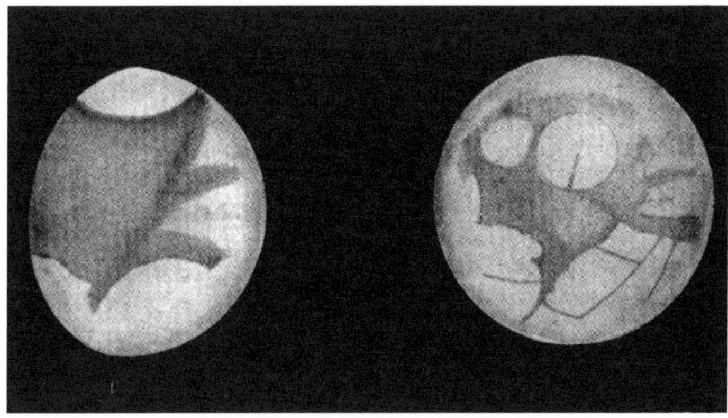

Fig. 67. Two Views of Mars, showing Seasonal Changes
The left-hand picture is a winter view; in the right-hand one the same portion of the planet during its summer is shown. The pole-cap seen during winter has disappeared by summer, and other features have also developed.
Drawn by P. Lowell, Lowell Observatory

Mars looks quite different in summer and in winter. This is shown in Figs. 67 and 68. In the former are two drawings of the planet, one made during its winter and the other during its summer. In Fig. 68 are photographs taken at that season on Mars which corresponds to our May and at other times on to the Martian summer. The change in the appearance of the surface is very clear. The great pole-cap, which is supposed to be ice, melts and almost disappears, while the markings at the equator grow in width and become darker.

The Canals of Mars

Within the last fifty years astronomers have detected some fine lines on the surface of Mars, which have been called *canals*. In Fig. 69 are four views of the planet, from drawings by Dr Percival Lowell. The faint straight lines are the canals. The lighter portions of the surface shown in these views appear reddish in the telescope, and the darker portions appear greenish. The former are supposed to be desert, and the latter vegetation of some sort. Some people think these straight lines are real canals which have been built by intelligent persons on Mars, but the majority of astronomers are very doubtful about that.

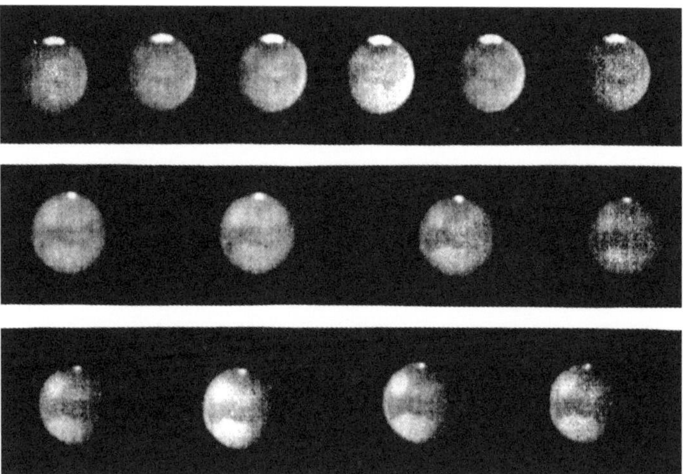

FIG. 68. PHOTOGRAPHS OF MARS, SHOWING CHANGES IN SPRING AND SUMMER
Clear photographs of Mars on a scale large enough to reveal details on the surface are hard to secure, as the image on the photographic plate is so small; but distinct changes produced as the Martian year advanced are shown in these pictures.
Photographs by E. C. Slither, Lowell Observatory

The canals can be detected best by observation with the telescope. They are too fine and faint to be photographed satisfactorily. An interesting comparison of photographs with a drawing made at the same time is given in Fig. 70. In this figure 1, 2, and 3 are photographs taken in the regular way – that is, with a camera at the end of a telescope. At the same time the drawing, 4, was made. Then this drawing was placed at such a distance from the camera that when photographed (through the telescope, of course) the image on the plate was of just the same size as the image of the planet in the sky. The picture thus taken is 5. It is precisely like the photographs 1, 2, and 3.

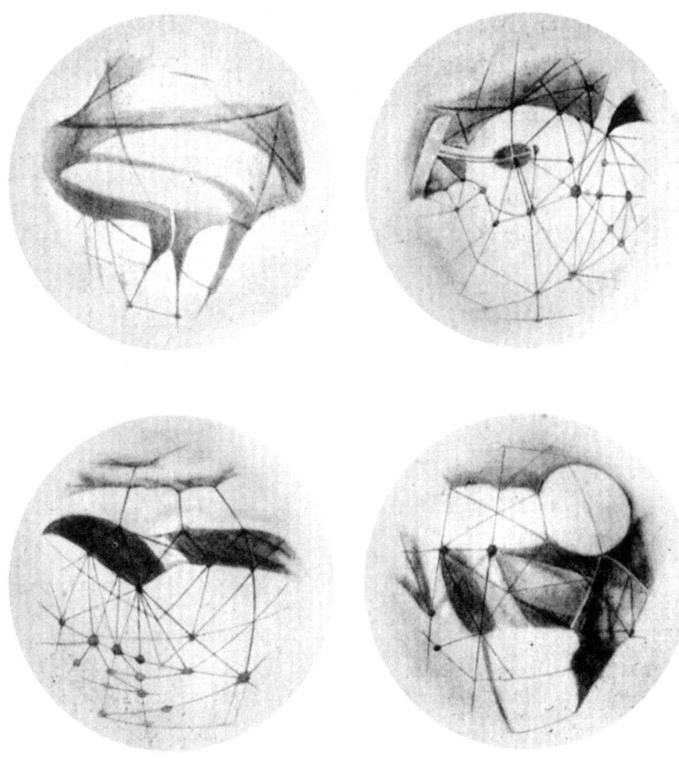

FIG. 69. FOUR DRAWINGS OF MARS (LOWELL)
These pictures exhibit the four 'sides ' of Mars as it rotates on its axis.
The features shown appear as the planet rotates 90° on its axis.
The rotation period of Mars is 24 hr. 37 min.
Redrawn by F. S. Smith

The Atmosphere of Mars

Since the Very surface of Mars can be seen clearly, it is evident that its atmosphere must be rare and transparent, not dense and opaque like that of Venus. A close watch on the planet, extending over many years, has revealed an occasional thin cloud, but the general opinion of observers has been that there is almost continuous fine weather on Mars – with no disturbances in the atmosphere such as produce storms on the earth.

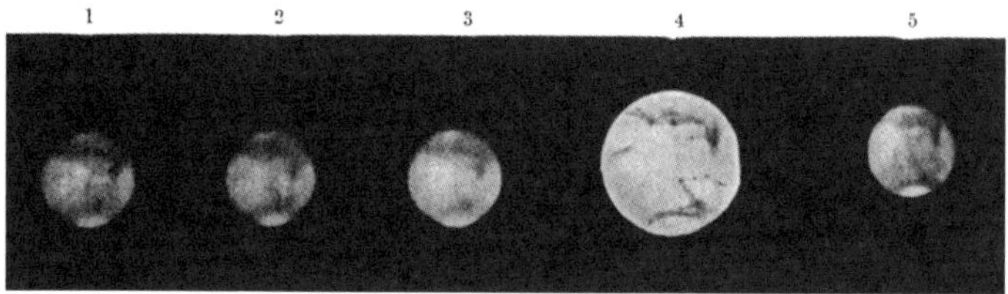

Fig. 70. Photographs of Mars, and a Drawing made at the Same Time
Even when Mars is nearest to the earth, its image on the photographic plate is small, and it is impossible to record in this way fine details which can be seen through the telescope. In this illustration the first three pictures are photographs, the fourth is a drawing made at the time, and the fifth is a photograph of this drawing when It was placed so far off that it appeared in the telescope to be of the same size as the actual planet did. It looks very like the direct photographs of the planet.
Photographs be E. C. Slither, Lowell Observatory

In 1924 and again in 1926, however, striking evidence of an atmosphere was obtained – and in a very neat manner. For some years Professor W. H. Wright of the Lick Observatory in California has been studying how best to photograph distant objects on the earth by means of a telescope. He found that the most satisfactory pictures were obtained when he placed a deep red filter over the photographic plate. In this way only red light was used in producing the image on the negative. If he used a blue or a violet filter, or no filter at all, many of the details were not recorded.

This is well illustrated in the two pictures in Fig. 71. They were taken from the summit of Mount Hamilton, on which is the Lick Observatory. The town of San Jose is thirteen and a half miles away in the famous Santa Clara valley. The right-hand picture was taken with red light. In it you can see clearly the town as well as the trees and roads in the valley. The left-hand picture was taken with

FIG. 71. PHOTOGRAPHS OF A DISTANT LANDSCAPE TAKEN WITH VIOLET AND RED LIGHT
On placing it violet filter before the photographic plate, and thus allowing only violet light to reach it, the left-hand picture was obtained; with a red filter the right-hand picture was secured. These are pictures of the city of San José, thirteen miles away from the summit of Mount Hamilton, California, where the Lick Observatory is located.
Photographs by Wright, Lick Observatory

violet light[1]. In it you see only that part of the mountain near the observatory, the distant valley being only a blur.

You remember, of course, that white light contains rays of all colours. It seems that the red rays can pass freely through the atmosphere, while the other rays are absorbed by it.

When Mars came especially close to the earth in 1924 the thought occurred to Professor Wright to photograph it with red and with violet light. The results were surprising.

Some of the photographs which he obtained are shown in Figs. 72 and 73. In each of these A is the picture taken with violet light, B that with red light. You see details with red light which are entirely absent from the pictures with violet light.

There is something else to be observed. The violet pictures are distinctly larger than the red ones.

Why is this? The explanation offered is that there is an atmosphere about the planet, and when the light from the sun reaches it the

[1]Actually these photographs were taken with infra-red and ultra-violet light – that is, with radiations just beyond the red end and the violent end of the spectrum.

violet rays are unable to pass down through the atmosphere and get to the surface, but are turned back by the atmosphere, and when

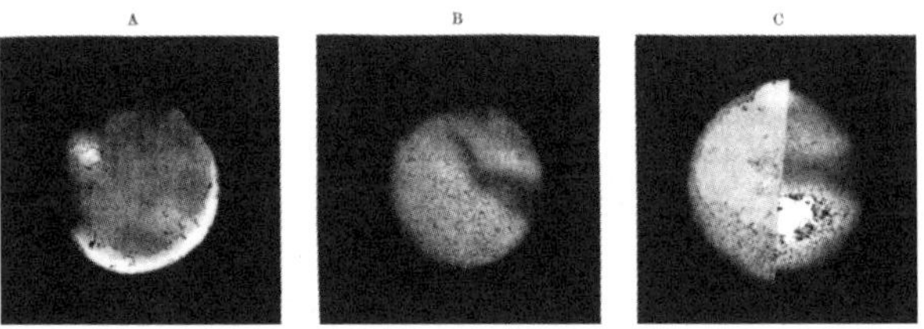

FIG. 72. PHOTOGRAPHS OF MARS TAKEN WITH VIOLET AND RED LIGHT
A, with violet light; B, with red light; C, half of much image placed together showing that the violet image is distinctly the larger. From this it is concluded that Mars has an extensive atmosphere.
Photographs by Wright, Lick Observatory

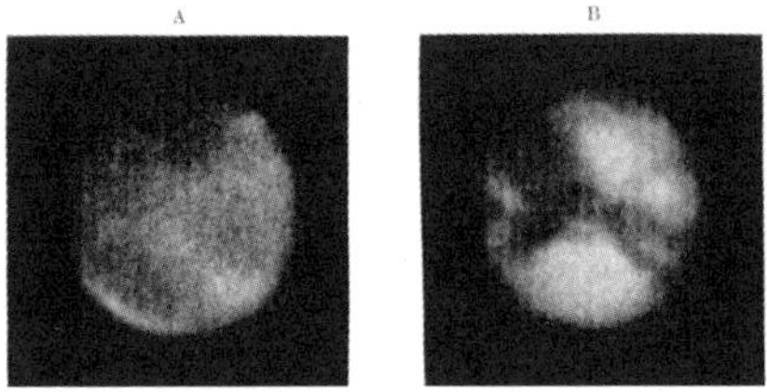

FIG. 73. FURTHER PHOTOGRAPHS WITH VIOLET AND RED LIGHT
The red image (B) is smaller, and shows more detail than the violet image (A).
Photographs by Wright, Lick Observatory

they arrive at the camera they simply give a photograph of the atmosphere, not of the surface of the planet. On the other hand, the red rays pass right through the atmosphere to the surface, and then come back through it. When they reach the camera they give a picture of the surface.

On placing one-half of a violet image against one-half of a red image (Fig. 72, C) it is clearly seen that the former is the larger. Indeed, by measurement the radius of the violet image is found to be about 5.7 per cent. greater than that of the red image. Now the red image seems to be a picture of the actual surface of the planet, and the extra 5.7 per cent. must be due to the atmosphere surrounding it.

FIG. 74. THE MARTIANS, ACCORDING TO MR H. G. WELLS
In Mr Wells's highly imaginative story *The War of the Worlds* the inhabitants of Mars are described as having their heads over-developed and enlarged but their bodies reduced to slender tentacles.
Drawn by F. S. Smith, after Dudouyt

The radius of the planet is about 2100 miles, and 5.7 of this is about 120 miles, which is, therefore, the height of the atmosphere.

Note the peculiar white 'excrescence' on Fig. 72, A. This is a specially large cloud. There is no trace of it in B. In Fig. 73, B, there is a white spot near the left edge of the picture of the planet and half-way down from the top[1]. This also is a cloud, it would seem, which in this instance is not seen in A.

Thus what began as simple amusement in amateur photography led to important results in astronomy.

[1]Turn the book until what is now the right side of the picture is at the top. The dark markings on the planet will look like a deer's head. The white spot is at the deer's throat.

Is Mars Habitable?

As has already been remarked, the winters on Mars must be extremely cold. How about the summers? Within the last few years the instrument known as a thermo-couple has been made so delicate and sensitive that it can actually measure the heat which Mars radiates to the earth, and from this heat calculations of the temperature of different portions of the planet's surface can be made. It is found that the temperature of the brightly illuminated surface is not unlike that of a cool day on the earth with the temperature ranging from 45° to 65°F. It is quite possible that some vegetable, and perhaps some animal, life may exist on Mars.

Some years ago Mr H. G. Wells published a very clever story entitled *The War of the Worlds*. In it he described an imaginary attack made upon the earth by people from Mars. The author represented the inhabitants of Mars as very advanced intellectually, their brains having expanded so that their heads were about 4 feet in diameter. Their bodies, however, had shrunk almost to nothing, and their legs were little more than slender, whip-like tentacles.

Of course, it is hardly necessary to remark that Mr Wells's story is purely fanciful, but it is told in a realistic and thrilling way.

A French artist, Monsieur Dudouyt, has drawn some pictures of the Martians according to Mr Wells's conception of them (Fig. 74). Notice that two of these beings are standing up, while a third one is sitting on the ground, unable to get up. Now the force of gravity, which tries to pull all bodies down toward the centre of the earth, or any other planet, is much smaller on Mars than on the earth. Indeed, it is only about two-fifths as great. A body which weighs 100 pounds when tested by a spring balance on the earth would, if taken to Mars and tested in the same manner, weigh only about 40 pounds. If a person from the earth could visit Mars he would find it very easy to rise up after lying down, and if he could jump 10 feet on the earth he would be able to jump 25 on Mars. But it would be quite different if a person from Mars came to the earth. He would probably find it impossible to get on his feet from lying or sitting down. In the story the Martians are represented as finding it difficult to move about the earth; and the one shown in the picture sitting on the ground finds it very difficult, if not impossible, to get up.

The Moons of Mars

Mars has two small moons (Fig. 75). They were discovered in 1877 by an astronomer named Asaph Hall. In the summer of that year Mars came very close to the earth, though not so near as in 1924, and a telescope with a diameter of 26 inches, the largest in the world at that time, had recently been erected at Washington, the capital of the United States. Night after is night in August Mr Hall searched the sky near to the planet to see if he could find any moons revolving about it. He was on the point of giving up, but his wife urged him to continue a little longer. He did so, and on the night of August 11 he discovered one moon, and six days later a second.

These moons are named Deimos and Phobos, which are the names given by the poet Homer to the horses which drew the chariot of Mars, the god of war. They are only a few miles in diameter, and give out very little light. They are very close to the planet, and revolve about it very rapidly – Phobos, the nearer one, in seven hours and thirty-nine minutes; and Deimos, the farther, in thirty hours and eighteen minutes. Just think of it – Phobos revolves three times about Mars while the planet rotates once on its axis. It actually rises in the *west* and rushes over and sets in the *east*. No other satellite behaves like this. Our moon takes four weeks to go round the earth.

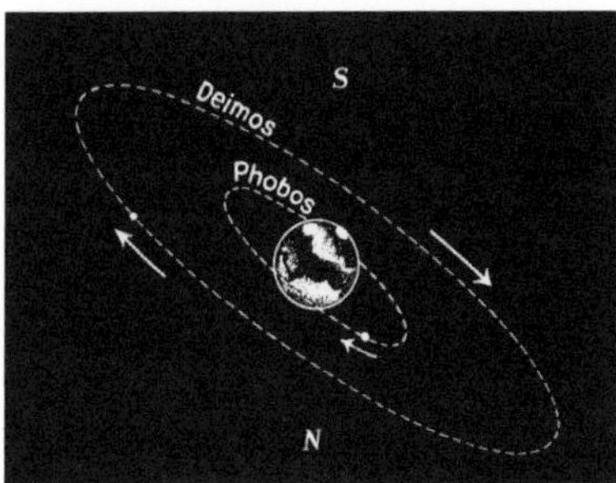

FIG. 75. ORBITS OF THE MOONS OF MARS

The inner moon revolves in 7 hr. 39 min., the outer in 30 hr. 18 min., while Mars rotates on its axis in 24hr. 37 mm. Thus the inner moon revolves three times in a Martian day. It moves so fast that it rises in the west and sets in the east.

CHAPTER VII

JUPITER—SATURN—URANUS—NEPTUNE

Jupiter

HUS far we have been learning about the four inner planets – those members of the sun's family which are always found comparatively near the great body about which they move. They are all comparatively small and composed of heavy, or dense, matter. Let us now travel farther out into space to see the other four members of the family as they pursue their courses at their immense distances from the great sun.

The first one we reach is the mighty Jupiter, the greatest of all the planets. Its distance from the sun is 5.2 times that of the earth, or 483 million miles, and it requires almost twelve of our years for it to complete a single revolution, although it speeds along continually at the rate of eight miles a second.

Jupiter as seen in a Telescope

Jupiter is a very charming sight through a small telescope as it hangs up there in the sky (Fig. 76). In addition to the big disc of the planet, four sparkling little stars are to be seen in a straight line drawn through the planet, like pearls on a stretched string. These are satellites, or moons. They were the first objects in the sky discovered by Galileo in 1610 with his newly invented telescope. They are just beyond the power of the naked eye, but can be seen with a good field – glass if it is held very steady.

FIG. 76. JUPITER AS SEEN IN A SMALL TELESCOPE
The four moons are seen in a row. Jupiter has five more moons, but they are so small that a large telescope is required to see them.

Through a larger telescope we see interesting bands and other markings on the planet (Fig. 77), and we also recognize that Jupiter is not a perfect sphere, but is decidedly flattened at the poles. This is clearly shown in the drawing. Actual measurement reveals the fact that, while the equatorial diameter is about 89,000 miles, the polar diameter is only 83,000 miles.

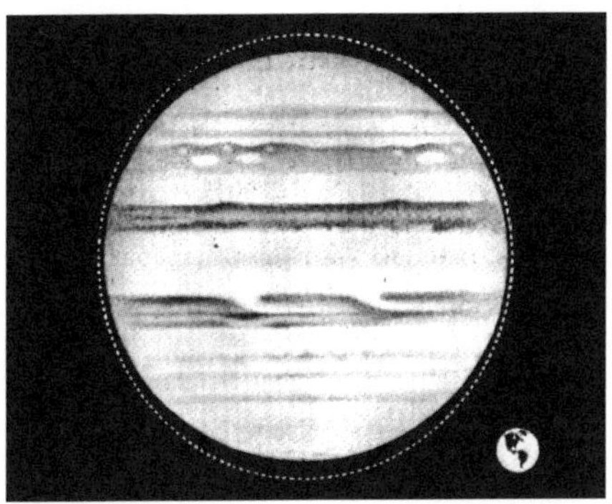

FIG. 77. JUPITER AS SEEN IN A LARGE TELESCOPE
In a large telescope curious details of the bands are revealed, and the brown and olive tints of the surface present a fine picture. The features change completely in two or three hours, as the planet rotates on its axis in less than ten hours. The flattening at the poles is also very evident, as shown in this drawing.
Drawn by Keeler, Lick Observatory

Why is Jupiter so flattened? We should suspect that it must spin on its axis very rapidly, and such is the case. It actually rotates once in nine hours and fifty-five minutes. A point on its equator travels over 28,000 miles per hour!

The Mass of Jupiter

Jupiter contains enough material to make 317 earths, but it is not so dense as the earth – indeed, only about one-quarter as dense. This is about the same as the density of the sun, and some people have surmised that Jupiter must be very hot. It does not, however, give out any light of its own; we see it only as it is illuminated by the sun's rays.

Moreover, recent measurements have shown that the temperature of its surface is very low.

The Motion of Jupiter's Satellites

A pleasing exercise for a person with a small telescope is to watch the four satellites. They are numbered I, II, III, IV in order of their distance from the planet. They are all larger than our moon, and their times of revolution are forty-two hours, three and a half days, seven days, and sixteen and a half days respectively. It is very interesting to follow their motions as they move about the giant planet. Sometimes they pass in front of it and cast a shadow on its face. The satellite and its shadow slowly move across the face of the planet, and apparently drop off at its edge. This is illustrated in Fig. 78, which shows two

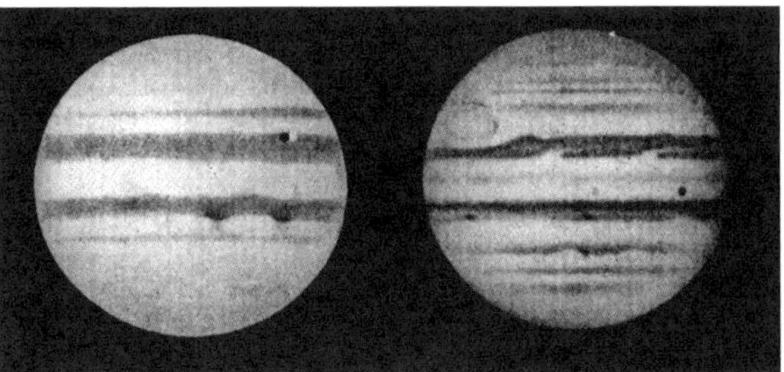

Drawn by J. H. Bridger, Croydon October 20, 1916, 11.05 P.M.

Drawn by J. E. Keeler, Lick Observatory August 30, 1890, 11.09 P.M. (*Pacific Time*)

FIG. 78. TWO DRAWINGS OF JUPITER, SHOWING A MOON AND ITS SHADOW IN TRANSIT

In each picture the satellite and its shadow are seen crossing over in front of the planet. The dark, round spot is the shadow upon the planet's surface cast by the satellite. In the left-hand picture the satellite is close to its shadow, both being well seen against a dark band of the planet. In these drawings the planet is shown as it appears in an ordinary telescope, and so the top is the south. In the telescope the satellite and its shadow enter at the right edge and move to the left across the disc. At the time this drawing was made the satellite had been on the disc sixteen minutes mid its shadow twenty-two minutes.

In the right-hand picture the dark shadow is clearly seen at the right. The satellite itself is slightly to the right of the centre of the disc. There is little contrast between it and the bright band behind it, and so it is not conspicuous. In this case the satellite had entered upon the disc an hour and six minutes, and the shadow twenty-one minutes, before the drawing was made.

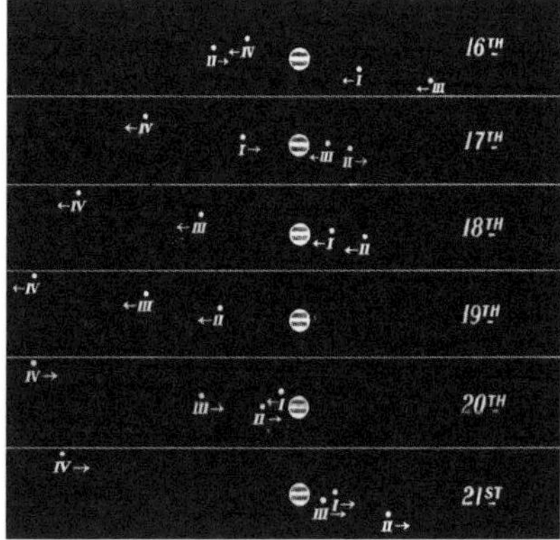

FIG. 79. JUPITER AND ITS SATELLITES ON SIX SUCCESSIVE DAYS
The central object is the planet, and the dots represent the four large satellites. Their distance from the planet and their periods of revolution about it are as follows: No. I, 261,800 miles; 42½ hr. No. II, 416,600 miles; 3 d. 13 hr. No. III, 664,200 miles; 7 d. 4 hr. No. IV, 1,168,700 miles; 16 d. 17 hr. In the diagram the positions are shown at 1 A.M. on July 16-21, 1926. The arrow indicates the direction in which the satellite is moving. It is interesting to follow their motions from day to day.

drawings of Jupiter. In the left-hand one a satellite and its dark shadow are seen close together as they move across the face. In the right-hand picture the dark round shadow is easily seen near the right-hand side, being in front of a bright band on the planet. The satellite itself, which produces the shadow, directly to the left of the shadow and a little to the right of the centre of the disc. You can hardly see it, for it is a bright object and is in front of a bright band.

In this picture notice also an elliptical object above the equatorial band and near the left edge of the disc. This is what was known as the *great red spot*. It was first noted in 1878, and was very conspicuous for several years. When at its best it was 30,000 miles long and about 7000 wide. Now it is gone, but the hollow where it lay can still be seen.

In Fig. 79 are given the positions of the satellites at 1 A.M. on six days in July 1926. The arrow indicates the direction in which the satellite is moving, and we can easily follow the motions of each satellite from one day to the next.

Photographs of Jupiter

Some excellent photographs of Jupiter have been taken, even though when nearest the earth it is more than 370 million miles away. In Fig. 80 are two photographs taken at the Lowell Observatory, one in 1915 and the other in 1917. The bands and some other details on the surface are well shown. At the upper right edge of the left-hand picture and above the equatorial band is seen the hollow where the great red spot was.

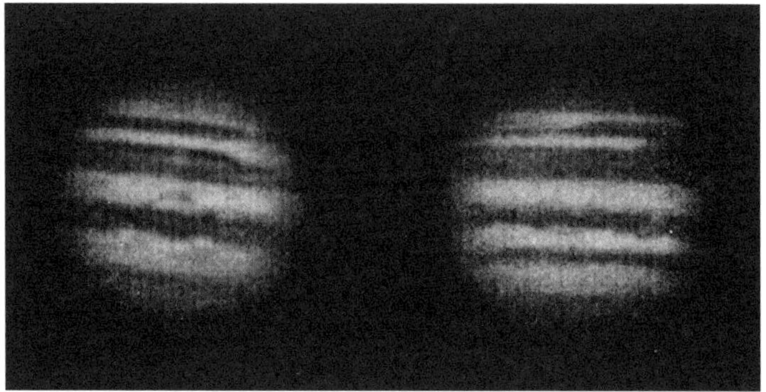

FIG. 80. PHOTOGRAPHS OF JUPITER
In these photographs, taken in different years, the changes in the system of belts are shown almost as clearly as in a good drawing made with a large telescope.
Photographs by E. C. Slither, Lowell Observatory

Professor Wright has also taken some photographs of Jupiter in ultra-violet and infra-red light. Two of them, the second one taken after an interval of an hour and sixteen minutes, are shown in Fig. 81. As in the case of Mars, the red image is the smaller of the two, and it shows some features which do not appear on the violet image. In the interval between the two photographs the planet made one-eighth of a rotation on its axis – the rotary motion being from right to left, as seen in the pictures. The hollow of the great red spot is just coming on in the violet image; in the other it is near the centre of the disc, but it is invisible.

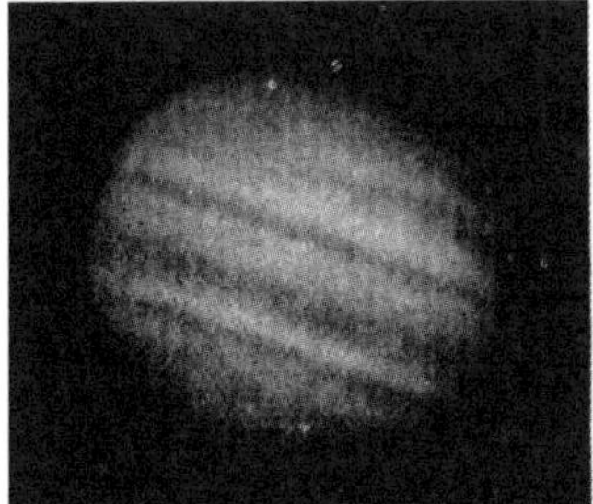

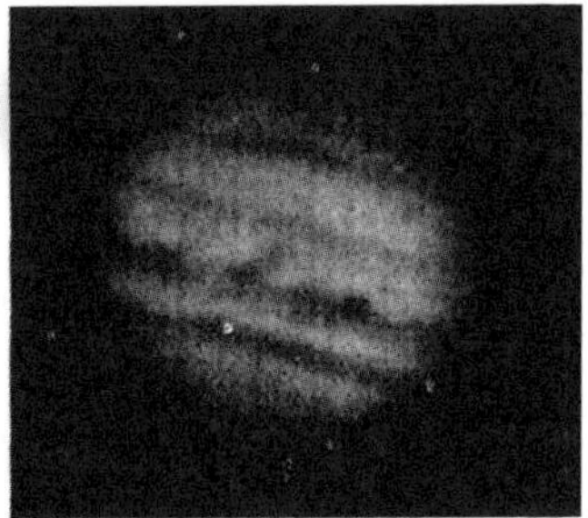

FIG. 81. PHOTOGRAPHS OF JUPITER TAKEN WITH VIOLET AND RED LIGHT
The left-hand picture was taken with ultra-violet light, the right with infra-red. The former was exposed at 3.9 A.M., October 18, 1926; the latter, 1 hr. 16 min. later. Between exposures the planet rotated 45°. The rotary motion is from right to left. The hollow where the old Red Spot formerly lay is just appearing on the left image; it is near the centre of the right image, but is invisible. Note the greater diameter of the ultra-violet image; also the flattening of the planet.
Photographs by Wright, Lick Observatory

Saturn

The next member of the solar system as we proceed outward from the sun is the most remarkable of all the family. It is the planet Saturn (Fig. 82). At nine and a half times the earth's distance, or 880 million miles, from the sun it moves along its path at the rate of six miles per second and completes a revolution in twenty-nine and a half years.

Its great belted globe is surrounded by a wonderful set of rings, and as it moves majestically along its appointed way it is attended by ten satellites which for ever revolve about it, some quite close, others a long way off, as though they were a bodyguard to protect it from enemies lurking in space.

The equatorial diameter of the ball is 74,100 miles, and the polar diameter is about 7800 miles less, or 66,300 miles. Thus it is much flattened at the poles, and we should expect it to turn about its axis rapidly. And such is the case. To make one rotation requires only ten hours and fourteen minutes.

The matter of which Saturn is composed is about as heavy as dry oak or maple-wood, and so the planet would float on water if a large enough ocean could be found on which to launch it. All the other planets would sink.

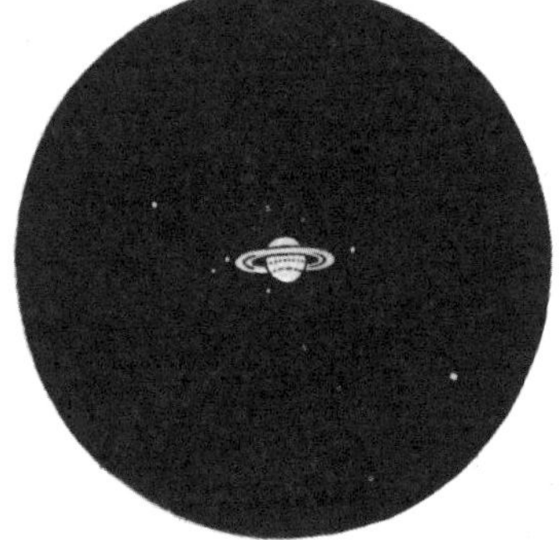

FIG. 82. SATURN AND ITS SATELLITES AS SEEN IN A SMALL TELESCOPE
Saturn, encircled by its remarkable rings and accompanied by its family of moons, is a charming object in the telescope.

The Rings of Saturn

There are three rings in the system, as is clearly illustrated in Fig. 83. The outer one is separated by a dark space from the middle one, which is wider and brighter than the first, and the innermost ring is much fainter than either of the others.

The outside diameter of the outer ring is 171,000 miles, while its thickness is probably not greater than 10 miles. If you should take a sheet of the thinnest tissue-paper and cut out a circular disc 17 inches in diameter to represent the size of the rings, the thickness of the paper would represent their thickness.

What are these rings made of?

They are composed of small, separate bodies, like boulders or big stones, which are so close together that very little, if any, light can pass through between them. These bodies revolve about the ball of the planet as though they were a close-packed swarm of moons, each independent of the others.

Phases of the Rings

As the planet revolves about the sun the rings show different phases as illustrated in Fig. 84. When either the earth or the sun is in the plane of the rings they cannot be seen. This happens every fourteen and three-quarter years. They disappeared from our view in 1907 and 1921, and will be invisible again in the year 1936, as shown in the diagram.

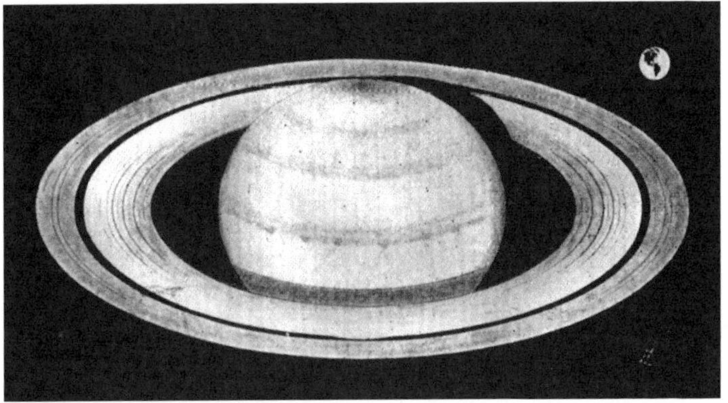

FIG. 83. A DRAWING OF SATURN

Note the great ball with its belts and the criss-cross markings at the equator and also the wonderful system of rings. The outer ring has an exterior diameter of 171,000 miles, and is about 10,000 miles wide. Cassini's division, between it and the middle ring, is 3000 miles wide. The middle ring has an outer diameter of 145,000 miles, and it is 16,000 miles in width. It is separated by a narrow division from the inner, or 'crape,' rain, which is 11,500 miles wide.

Drawn by Percival Lowell

Though Saturn when closest to the earth is still about 800 million miles away, some wonderful photographs of it have been taken. In Fig. 85 are reproduced three from negatives taken at the Lowell Observatory. They show the phases clearly.

To the naked eye Saturn gives a dull, steady light. Venus is so brilliant that no one need mistake it. Jupiter comes next in brightness and, though much inferior, is still brighter than any fixed star. Mars on rare occasions actually outshines Jupiter, but usually it is much fainter. Its reddish light, however, renders it easy to recognize. Saturn appears as a first-magnitude star – about as bright as Arcturus or Procyon. But it has a yellow tinge and does not twinkle. Indeed, none of the planets, except Mercury, ever twinkle, unless they are very close to the horizon.

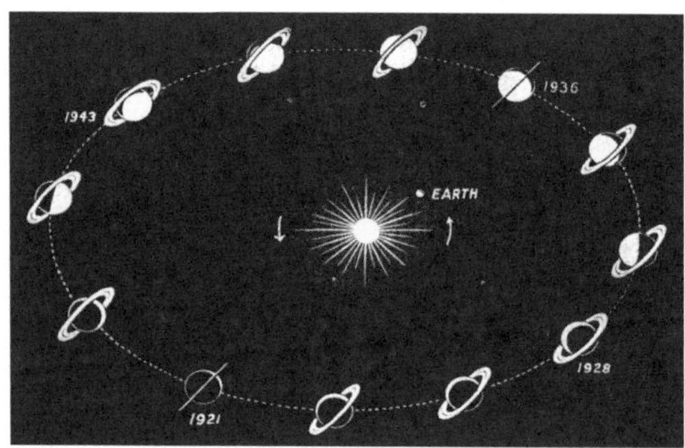

FIG. 84. WHY SATURN'S RINGS SHOW PHASES
As the planet revolves about the sun its rings are presented edgewise to the sun and the earth twice during its period of twenty-nine and a half years. At such times the rings are invisible, except perhaps in the largest telescopes. Years when this happens (also those when the rings appear opened widest) are indicated in the drawing.

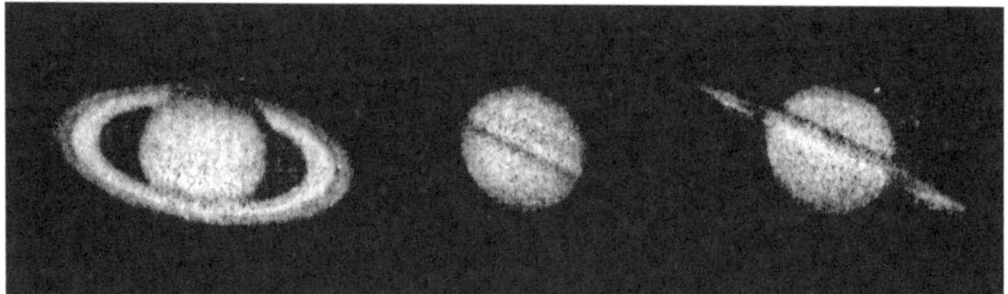

FIG. 85 PHOTOGRAPHS OF SATURN, SHOWING PHASES OF THE RINGS
These beautiful photographs – taken in February 1916, April 1921, and May 1922 – show Saturn with its rings most spread out, when turned edge-on, and a year later, when slightly opened up. Note Cassini's division and the shadow on the rings cast by the ball.
Photographs by E.C. Slipher, Lowell Observatory

Uranus

For many centuries it was thought that Saturn was the most distant of all the planets, and that its orbit formed the outer-most boundary of the solar system; but a startling discovery made in 1781 showed that it was necessary to travel out twice as far as Saturn's orbit before reaching the limit of the sun's kingdom.

On March 13 of that year William Herschel, a professional musician living at Bath, in England, was studying the sky with a telescope which he himself had made. Herschel had found that in order to understand fully the theory of music he had to study algebra, geometry, and other branches of mathematics. From these he was led to study optics, or the science of light, and the construction of the telescope; and, having used a small telescope to observe some of the objects in the sky, he desired to obtain a larger one. He inquired the price, but it was more than he could pay, and he determined to make one himself.

Two Kinds of Telescopes

There are two distinct kinds of telescopes. In one a lens is used to collect the light from the celestial object; in the other a mirror performs this service. The former is called a refractor; it is the kind you ordinarily see. The latter is called a reflector, because the light is reflected from the surface of the mirror. Now it is much easier to make a mirror than a lens, and it was a mirror telescope (or reflector) which Herschel constructed. Indeed, it is quite a favourite exercise for an amateur astronomer to make his own telescope, and many good ones have thus been made. To grind and polish a mirror demands patience and care – and delicate skill too, if a good one is to be made.

Here is a telescope (Fig. 86) constructed by a Toronto lawyer, who is seen observing with it. The mirror is in the lower end of the tube, and the eyepiece, into which the astronomer looks, is at the other end. The tube in this case is of galvanized iron, and it is supported on

[1]The huge tube of this telescope is 30 feet long, and 7 feet 6 inches in diameter. The upper portion (23 feet in length) is a framework of steel. The great mirror is at the bottom of the enclosed portion. It is an immense disc of glass, 72 inches in diameter, 12 inches thick at the edge, and 11.1 inches at the centre, where a hole 10 inches in diameter is bored out of it. It weighs 4340 pounds. The small mirror is seen at the top of the tube. A spectrograph is attached at the lower end of the tube.

FIG. 86. AN AMATEUR AND HIS TELESCOPE
The mirror (not a lens) of the telescope is at the lower end of the sheet-metal tube, which is mounted on a wooden post with the assistance of iron pipes and iron weights. The maker (Mr A. R. Hassard, of Toronto) is observing at the eyepiece, which is at the upper end of the tube.

a wooden post. The largest telescopes in the world are constructed on this principle. At the Mount Wilson Observatory, in California, is one 100 inches in diameter. This is the greatest of all. The next in size[1] is at the Dominion Astrophysical Observatory, near Victoria, British Columbia (Fig. 87). It is 72 inches (6 feet) in diameter. These great instruments are triumphs of engineering skill. They are very massive and complicated, but they can be handled very easily, and they move with the precision of a watch.

In Fig. 88 is shown the greatest lens telescope in existence.

It is in the Yerkes Observatory, which is a part of the University of Chicago, and is located at Williams Bay, Wisconsin. The lens is 40 inches in diameter, and the telescope is about 65 feet long.

The telescope which Herschel was using on that day in 1781 was

Fig. 87. The 72-inch Reflecting Telescope of the Dominion Astro-physical Observatory, Victoria, British Columbia

FIG. 88. THE GREAT REFRACTING TELESCOPE OF THE YERKES OBSERVATORY
This is the largest lens telescope in existence. The diameter of the lens is 40 inches, and its focal length is 62 feet.
Photograph by Ross, Yerkes Observatory

6¼ inches in diameter and 7 feet long. It was his custom to rush home after a practice or a concert, and spend the rest of the night in observing the heavens. He was usually accompanied by his sister Caroline, who would record what he observed.

How Uranus was discovered

As he was examining star after star in the constellation Gemini (the Twins) he came upon one which looked peculiar. In a good telescope a star appears simply as a bright point, but to his sharp eye this one looked larger – like a tiny disc. He tried eyepieces of higher power, and the disc looked larger. A fixed star does not behave in that way, and so he decided it could not be an ordinary star. A few days later he examined it again, and it had moved among the stars. He concluded that it was a comet, very far off and without a tail. The idea that it was a planet did not enter his mind.

After a few months it was proved to be a planet, and the name Uranus was at length given to it.

King George III knighted Herschel, gave him a pension, and made him the King's Astronomer. He constructed many larger and better telescopes, which were famous all over Europe, and he devoted much time to observing the heavens. Indeed, he became the greatest observational astronomer of all time.

FIG. 89. RELATIVE SIZES OF THE PLANET URANUS AND THE EARTH
The diameters are 32,400 miles and 7918 miles -*i.e.*, in the ratio 4.1 : 1.

Uranus and the Earth

Here is a drawing of Herschel's planet and the earth on the same scale (Fig. 89). The planet is 1800 million miles from the sun, and its orbit is so large that, though it moves with a speed of four and a third miles per second, it requires eighty-four years to complete its circuit of the sun. Its diameter is about 32,000 miles, and so its volume is sixty-six times that of the earth; but the material in it is not nearly so dense as that of the earth. It 'weighs' only fifteen times as much as the earth.

Uranus has four small satellites which can be seen only in large telescopes; but though they are so tiny and so far away they have been photographed. The four small dots to be seen in the picture (Fig. 90) are the satellites, and their names are given in the key below.

FIG. 90. A PHOTOGRAPH OF URANUS AND ITS SATELLITES
The moons of Uranus are very small and can be seen only in a large telescope, but they have been successfully photographed. The names given to them are to be found in the key.
Photograph by Lapland, Lowell Observatory

Neptune

Uranus is not the last planet in our solar system. There is at least one more. Its name is Neptune. Uranus was met with almost accidentally; Neptune was discovered by the mathematician in his study.

After Uranus had been observed for some months astronomers were able to calculate the size and position of its orbit, and it then became possible to predict where the planet would be found day after day. As the years went by, however, it did not occupy precisely the positions assigned to it. It did not wander very far from its computed path, but yet far enough to make the astronomers wonder, and at last they decided that there must be some far-distant body which was luring the lonely Uranus from the orbit assigned to it.

Discovered by Mathematicians

Two mathematical astronomers, Adams in England and Leverrier in France, both without any knowledge of the other, undertook to calculate just where the unseen trouble-maker was. It was a very difficult problem, but they both solved it. They stated just where in the sky to look for the stranger, and on September 23, 1846, an astronomer in Berlin pointed his telescope to the place and found it. The discovery caused a great sensation.

Neptune is 2800 million miles from the sun. It moves along its orbit at the rate of three and a third miles per second, and requires 164 years to perform a revolution. Thus it has completed only half a revolution since it was discovered. It requires field-glasses or a small telescope to see it.

Fig. 91 shows a drawing of Neptune and the earth. Neptune's diameter is about 31,000 miles. It is large enough to contain sixty earths, but the matter in it, or its mass, is sufficient to provide only seventy earths.

Discovery of Neptune's Satellite

Within a month of the discovery of the planet a little moon was detected revolving about it. Though they are at such an enormous

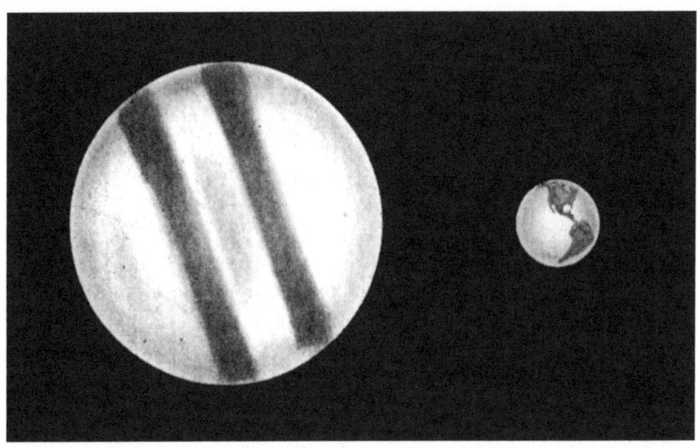

FIG. 91. RELATIVE SIZES OF THE PLANET NEPTUNE AND THE EARTH
The diameters are 31,000 miles and 7918 miles- *i.e.*, in the ratio 3-9 : 1.

distance from the earth, it has been found possible to take a photograph of the planet and its solitary satellite. In Fig. 92 the bright dot in the centre is the planet, and the little dot beside it is the satellite. It is about as far from Neptune our moon is from the earth.

Fig. 92. A Photograph of Neptune and its Satellite
The bright dot is the planet, and the faint object just left of it is its moon, which, at a distance of 219,800 miles from Neptune, revolves about the planet in 5 d. 21 hr.
Photograph by Lampland, Lowell Observatory

CHAPTER VIII

ASTEROIDS – THE NEBULAR HYPOTHESIS – COMET – METEORS

The Asteroids or Planetoids

We have now come to the frontier of the solar system, but there are one or two things which we must consider before we leave it.

Let us go back to the wide space between the orbits of Mars and Jupiter. When we look at the spaces between the orbits of the planets (Fig. 24) this one seems unnaturally large; and during many years the belief continually grew that there must be some planets, probably small, revolving about the sun in this wide space. At last a number of astronomers united themselves into a society which was jokingly named the Celestial Police, their intention being to hunt for those members of the sun's family which were hiding from view. The first culprit was arrested, quite accidentally, on January 1, 1801; and in the century and a quarter since then more than 1500 have been discovered, latterly by photography. In 1926 115 new ones were located, and in 1927 96 more.

The method is as follows. The astronomer points his camera to a part of the sky where he suspects that one of these planetoids, or asteroids, may be, and he keeps his camera moving so that its movement exactly corresponds to the motion of the sky. The stars then appear on his plate as round dots. Now a planet is continually moving among the stars, and if there is a planet among the stars photographed it makes a trail on the plate. In Fig. 93 there are two of these short trails recording the presence of two asteroids. The exposure was one hour.

A Family with Fifteen Hundred Members

The asteroid family has now grown so large that it requires much time and labour to keep a record of their doings. They are mostly very small bodies only a few miles in diameter. There is but one in the entire family which is bright enough to be seen without a telescope. Its name is Vesta, and it was discovered in 1807. Its diameter is perhaps 300 miles, but many of them are probably not more than three or four miles in diameter. Indeed, some are, in all likelihood, just irregular masses of rock.

FIG. 93. A PHOTOGRAPH SHOWING TWO ASTEROID TRAILS
The asteroids, or little planets, are continually moving in relation to the stars. In this photograph the camera was moved so as to follow the stars, the images of which are therefore round dots, while the asteroids make trails on the plate. Two are seen, not far from the centre of the photograph.
Photograph by by Barnard, Yerkes Observator

The Past History of the Solar System

Before we go any further let us look back and make a general survey of the solar system. It will be useful to write down a few of the facts we have learned about it.

(1) The orbits of the planets are all nearly circular and in planes which nearly coincide; that is, the angles between the planes are very small.

It is usual to consider the plane containing the earth's orbit (which is the same as the plane of the ecliptic) as the standard plane of reference. All the other planes of the planets (not counting the planetoids) make very small angles with it.

(2) The planets all move in the same direction in their orbits.

(3) At the centre the sun, which controls the motions of the planets, rotates in the same direction as that in which the planets revolve, and its equator is nearly in the plane of the ecliptic.

(4) The rotations of the planets, as far as they have been certainly determined, are in the same direction as their revolutions in their orbits.

(5) The satellites of the planets, with very few exceptions, revolve about their planets in the same direction as the planets rotate; their orbital planes also nearly coincide with the ecliptic plane.

(6) The earth and the sun are composed of the same materials, and we naturally suppose that the other planets are also composed of the same substances, though some of them are in a different physical condition from the earth.

What thoughts naturally arise on pondering over these remarkable facts? First, surely, that all these bodies had a common origin!

Many people love to imagine, or to speculate, just how the solar system developed into its present state. From a study of its present condition they try to learn its past history.

How the Solar System came into being

The first fairly satisfactory explanation of the way this took place was the Nebular Hypothesis which was put forth more than a hundred years ago. It was imagined that in the long-distant past a vast quantity of the thinnest gas – in other words, a nebula – filled a space extending much beyond the orbit of the outermost planet. It was supposed, further, that this gas was at a high temperature and was, as a whole, rotating about an axis. What would happen?

In the course of ages the nebula would cool down and contract, and the smaller it became the faster it would rotate. At last the speed would become so great that, it is believed, the material farthest from the axis would break away, just as water flies off a motor-car wheel as it rotates. According to the theory, the matter which broke away came together by its own attractive force and formed the most distant planet.

The operation did not stop here. The nebula continued to contract and to increase its speed, and other portions broke away from the mass, going to form the other planets. The portion which was finally left became the sun.

This theory can account for some features of the solar system, but it fails in several important matters. In consequence it has been given up by most astronomers.

Another theory, quite different in nature, was proposed a few years ago. In this case we are asked to believe that originally our sun was a simple, isolated star without planets attending it, and that at a remote time another star in its motion through space happened to pass near the sun. When they were very close there would be a tremendous attraction between the two bodies, and the sun was drawn out of shape. This caused it to burst open, and great quantities of matter were thrown out from it. The visiting star, as it moved away, set the sun in rotation and caused the thrown-out matter to circulate about the sun. From this matter in the course of time the planets were formed.

The above is the briefest outline of these two theories, but the subject is full of difficulties and cannot be considered further here. It is well to remember, however, that these are theories which can never be proved. Usually the astronomer is able to build upon the solid foundation of accurate observation and exact calculation, and then his predictions can be depended upon.

Comets and Meteors

Comets and meteors may also be considered to belong to the solar system, since they revolve about the sun, but they do not stop with us always. Fig. 94 shows how a comet behaves. When first seen it is very faint; it is coming from far out in space, and is just near enough for us to recognize it. Then as it approaches the sun, as indicated by the arrow, it develops a tail and becomes larger and brighter. When it gets to the point where it is nearest to the sun it is said to be in *perihelion*, and its tail is usually longest. Then it moves off, always having its tail directed from the sun. It gets fainter and fainter and at last disappears, probably never to be seen again.

Sometimes, however, the comets do come back, in which case they may very properly be counted permanent members of the sun's family. The one that returns to the sun most frequently is called Encke's comet, having been named after an astronomer who studied its motions. A photograph of it is shown in Fig. 95. It is too faint to be seen with the naked eye, and, as you see, it has no tail at all. It returns once in every three-tenth years.

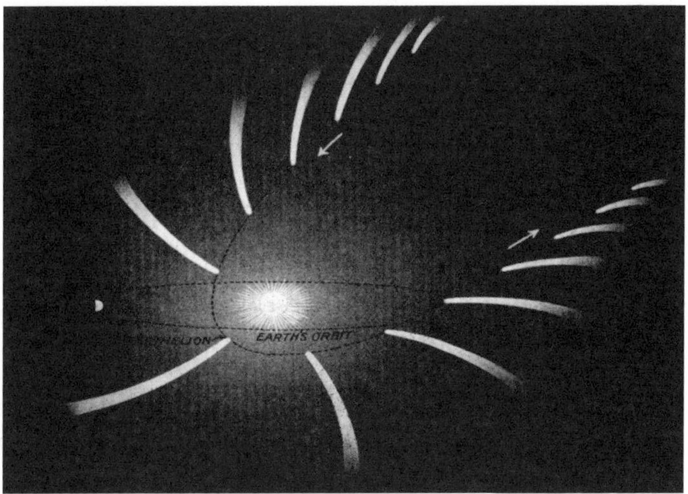

FIG. 94. THE ORBIT OF A COMET

This is the appearance of a great comet as it comes from the depths of space, approaches the sun, reaches perihelion and then sweeps off into space again. In its course it passes near the earth's orbit, and may come very close to the earth. On some occasions the earth has been completely enveloped in the tail of the comet.

Drawn by F. S. Smith

FIG. 95. A PHOTOGRAPH OF ENCKE'S COMET
This comet has no bright nucleus and no tail; it is only a faint, hazy patch in the sky, invisible to the naked eye. The exposure for this photograph was 1 hr. 53 min. The camera was kept pointed at the comet, which was moving among the stars, and so the latter made trails on the plate.
Photograph by Barnard, Yerkes Observatory

Their Strange Appearance

Encke's comet, you will agree, has nothing startling about its appearance, but some comets are remarkable objects. In Fig. 96 is shown one which was discovered by Brooks, an American astronomer, in 1911. Look

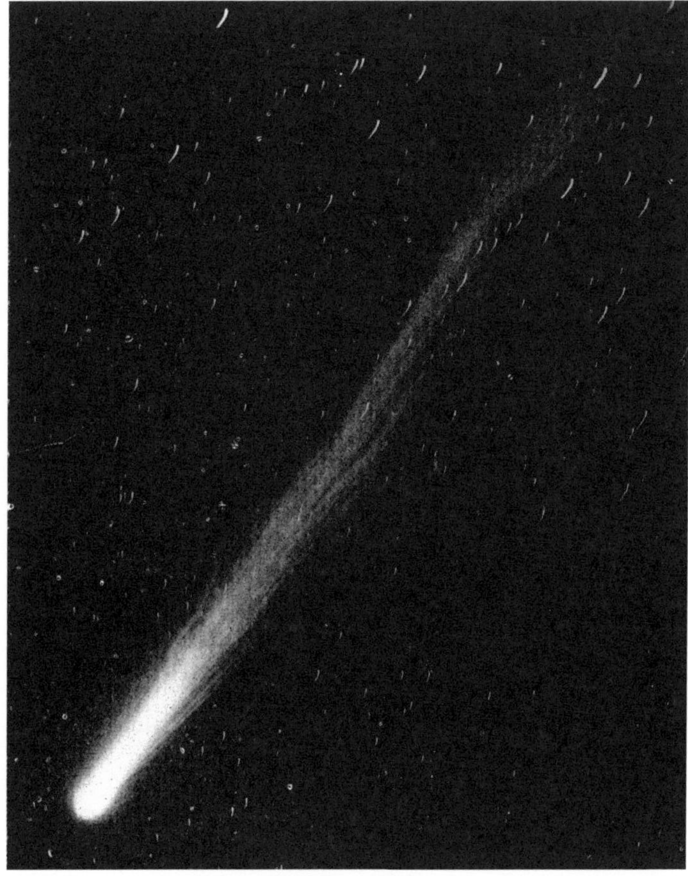

FIG. 96. A PHOTOGRAPH OF BROOKS' COMET
This comet when first seen had no tail, but as it approached the sun the tail developed rapidly, and in time the comet became a striking object with a tail 30° long. Note the delicate strands of the tail as they stream away. Length of exposure for this photograph, which was taken on October 23, 1911, 1 hr. 15 min.
Photograph by Barnard, Yerkes Observatory

at the wonderful tail as it streams away from the head of the comet. Another comet, discovered in 1908 by Morehouse (Fig. 97), exhibited some strange changes during the few months when it was visible. In the illustration you can see the many filmy ribbons of the tail as they are driven backward by the pressure exerted by the light from the sun. Indeed, at one place there is a great lump, as though the tail had been broken and a large portion were flying off into space.

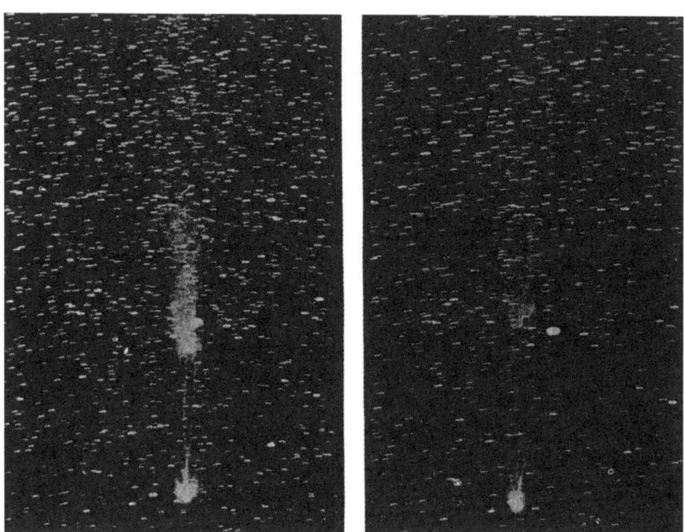

FIG. 97. THE ORBIT OF A COMET
This is the appearance of a great comet as it comes from the depths of space, approaches the sun, reaches perihelion and then sweeps off into space again. In its course it passes near the earth's orbit, and may come very close to the earth. On some occasions the earth has been completely enveloped in the tail of the comet.
Drawn by F. S. Smith

Halley's Comet

The most famous of all comets, however, is that named after the English astronomer Halley. It returns every seventy-six years. It came to the sun in 1910, and will certainly return in 1986. Some who read this page may live to hail it when it comes back to visit the solar system again. Fig. 98 is a photograph of this wonderful traveller taken on May 13, 1910. The exposure was one of thirty minutes. Notice the peculiar expansion in the tail. Also observe the bright spot near the bottom of the picture. That is the planet Venus, which happened to be in the same part of the sky as the comet at that time. Look also for a faint straight oblique streak at the right-hand side of the comet and one inch from its head. That was made by a shooting star. Thus in this photograph there are four kinds of heavenly bodies – a comet, a planet, a shooting star, and the fixed stars.

The Nature of Comets

There is still much that we do not know regarding the real nature of a comet. Its volume is often enormous. The head may be 100,000 miles or more in diameter and the tail 50,000,000 miles long, and yet the mass is very small – less than that of the smallest planet. That the matter in a comet is very thin is proved by the fact that a faint star viewed through the head of the comet does not seem to lose any of its brightness.

In the centre of the head is sometimes seen a bright, star-like point; this is known as the nucleus. It probably contains pieces of solid matter widely separated from one another. In a cubic mile there is, on the average, probably not more than half an ounce of solid material. As the comet approaches the sun the heat causes these solid bodies to give out gas accompanied by fine dust.

The tail consists of very fine particles which are expelled from the head and driven away by the sunlight, which exerts a pressure upon them. By comparing photographs taken at short intervals, as those in Fig. 97, it can be seen that the matter in the tail is moving away from the head. Indeed, in Halley's Comet a portion which started from the head was observed for over two weeks. When it was 800 miles from the head its speed was six-tenths of a mile per second, but when at a distance of 8,400,000 miles it was travelling at the tremendous speed of 57 miles per second. It is thus evident that a comet is continually losing its material, and, since there is no means that we know of by which it recovers any, it will in the end disappear.

FIG. 98. A PHOTOGRAPH OF HALLEY'S COMET
In this picture we see the records left by four diferent kinds of celestial objects. The great comet, with a portion of its tail apparently breaking out and rapidly flying away, is the chief feature. Then the bright round object at the right-hand side is the planet Venus. Also note a faint straight oblique streak about an inch above the comet's head: this is a shooting star. And the fixed stars cover the background of the sky. This picture was taken on May 13, 1910, and the exposure was thirty-six minutes.
Photograph by E. C. Slipher, Lowell Observatory

The spectroscope has shown that in a comet there is nitrogen gas, and also the poisonous gases carbon monoxide and cyanogen, but, though the earth has on several occasions passed through the tail of a comet, no ill effects were noticed.

Shooting Stars

Shooting stars such as that one recorded in the photograph (Fig. 98) are familiar to all. Usually there is a faint streak of light in the sky, which lasts but a moment and is gone.

These streaks are due to bits of matter, perhaps like grains of sand, which are rushing through space and happen to come into our atmosphere. Through the friction against the air they become white-hot and probably burn up.

Some persons wonder how a grain of sand can possibly continue moving through space with a speed of several miles per second. It seems natural for a great planet to do so, but not for a wee particle of stone. Remember, however, that out in free space, far away from the earth or any other planet, there is nothing to stop the motion, and so a body, whether large or small, travels forward unchecked.

A Great Fire-ball

Occasionally something bigger than usual comes along – perhaps a rock a foot or more in diameter – and as it ploughs its way through the atmosphere it makes a great display. In this case it is called a fire-ball; and a remarkable one is shown in the photograph reproduced in Fig. 99. An astronomer named Klepesta, at Prague in Czecho-Slovakia, was taking a long-exposure photograph of the stars on the night of September 12, 1923, when at 11 P.M. a brilliant fire-ball moved across the sky and left its mark upon his plate.

It fell slowly, like a burning sphere, the brightness continually increasing, and it was followed by a train of yellow luminous powder which lasted for about ten seconds. When the fire-ball, as it was seen projected against the sky, appeared to be near the great nebula in the constellation Andromeda (seen at the right-hand side of the picture) it exploded like a rocket and lighted up the whole country about.

FIG. 99. A PHOTOGRAPH OF A FIRE-BALL
Very seldom has such a brilliant fire-ball been photographed. The wider portions of the trail were probably produced by successive explosions in the meteor as it sped along through the air at the rate of thirty-seven miles a second. This photograph was taken at the National Observatory of Czecho-Slovakia.
Photograph by J. Klepesta, Prage, Czecho-Slovakia

You can see the double streak made by the two pieces.

Other persons also saw this wonderful meteor, and from all the observations made upon it it was possible to compute just where it was and how it was moving. When the body was first seen it was eighty-five miles above the surface of the earth, and after travelling fifty-five miles, at the rate of thirty-seven miles per second, it disappeared at a height of thirty-five miles above the earth.

Only seven and a half miles of its path is on the photograph, and at that time its height was fifty miles.

Fig. 100 shows another fine picture of a bright meteor, obtained by W. J. S. Lockyer, an English astronomer, while photographing the polar stars on November 16,1922.

Sometimes these bodies fall to the earth, in which case they are called meteorites. Many have been found, and can be seen in our museums. Some are almost solid iron, but generally they are of a stony nature.

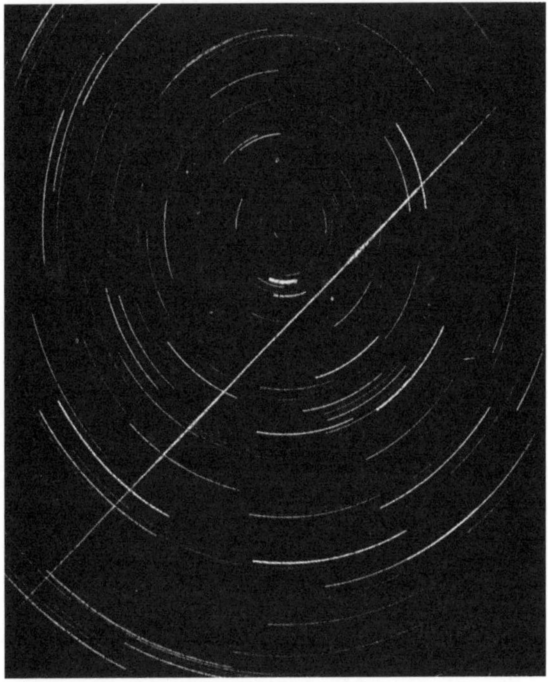

FIG. 100. A METEOR AMONG THE POLAR STARS
This interesting photograph was obtained at the Norman Lockyer Observatory, Sidmouth, Devon, on November 16, 1922. Exposure 2 hr. 14 min.
Photograph by W. J. S. Lockyer

PART III

THE UNIVERSE OF STARS

CHAPTER IX

THE STARS IN THEIR SEASONS

here is much yet to learn about the solar system, but we cannot longer delay. We must now turn from our own family affairs to look into the great celestial world beyond.

You are all accustomed to studying maps of the earth's surface, and you know that one map may cover only a small portion of the country, such as a county or a manufacturing district, while another may include an entire continent. So it is with maps of the sky. Some include large areas, while others take in single constellations or perhaps even a portion of one.

There is one thing which you must always remember, however. We view the earth from the outside, but the celestial sphere is seen from the inside. Its inner surface seems to be studded with stars.

The North Polar Stars

Let us now study some maps of the sky. Those which we shall take up first will cover large portions of the celestial sphere. Our first map (Fig. 101) contains the stars which you see when you face the north and look upward toward the Pole Star. On any clear night these stars can always be seen by a person in our middle northern latitude, but they are not always in just the same positions. They seem nailed to the sky and continually revolving about the pole of the sky, making a complete revolution in a day. Their positions vary with the seasons, too. In the autumn in the early evening the Plough or Big Dipper is seen below the Pole Star; in the spring at the same hour it is above it.

In Fig. 101 the stars are shown as you see them in the sky; on the opposite page is a key (Fig. 102) giving the names of the constellations and of the brighter stars. Round the circumference of the key are the names of the months. You notice that 'November' is at the top. The

FIG. 101. THE CIRCUMPOLAR STARS, AS SEEN IN THE SKY
This chart shows the positions of the North Polar Stars at about 9 P.M. in November.
Their positions six months later will be seen when the chart is turned upside down.
In the spring the Plough, or Dipper, is above the Pole in the early evening.

map as you are looking at it now shows the positions of the stars at 9 P.M. in November. To find their positions at the same hour in any other month, turn the map until that month is at the top.

The Great Bear and the Little Bear

The star very near the centre of the map is the Pole Star, usually called Polaris by the astronomer. Some distance below it is the Plough,

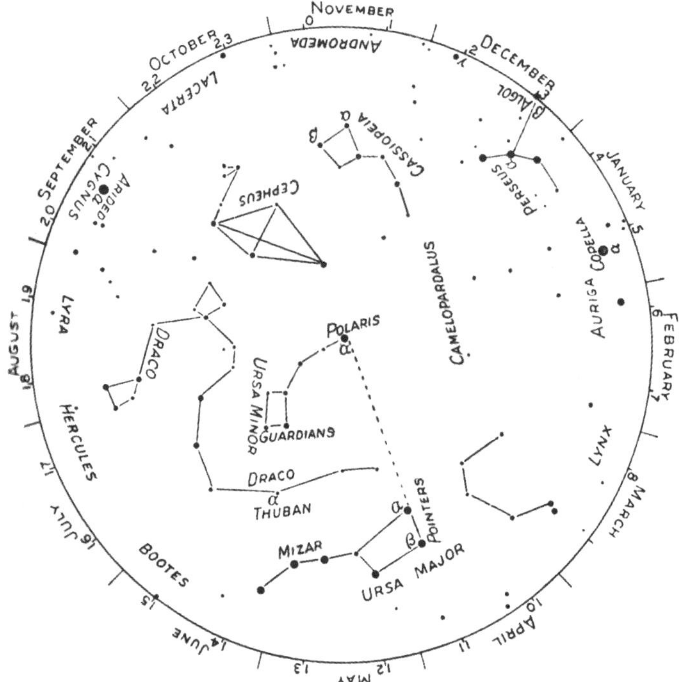

FIG. 102. KEY TO FIG. 101, GIVING NAMES OF THE CONSTELLATIONS
The positions of the stars at 9 P.M. during any month is given by holding this key so that the name of the month is uppermost.

or Dipper, which is part of the constellation Ursa Major (the Great Bear). The shape of these objects is traced out by seven bright stars. In star maps it is customary to denote each bright star by a letter of the Greek alphabet; but in this map only the first two letters, *a* (*alpha*) and *β* (*beta*), are used. *a* and *β* of the Great Bear are named the "Pointers." They are about 5° apart. If you draw a straight line from *β* through *a* and extend it about 30° beyond *a* it will come almost to the Pole Star.

The ancient Arabs gave the name Dubhe to *a* and the name Merak to *β*. They also called the star at the bend of the beam of the Plough, or the handle of the Dipper, Mizar. If you look sharply you will see near it another faint star. It bears the name Alcor.

The stars in the Plough are said to be of the second magnitude. There are about twenty stars in the sky which are decidedly brighter than these, and they are said to be of the first magnitude. After them come stars of the second, the third, and other magnitudes. The very faintest stars which you can see on a clear night out in the country when there is no moon in the sky are of the sixth magnitude. In the city you can hardly see anything below the fourth magnitude.

Before the telescope was invented (1608), of course, no stars fainter than the sixth magnitude were known, but since then much fainter ones have been shown to exist, and as telescopes have increased in size fainter and still fainter stars have been revealed. The giant instruments now in use will show by visual observation stars of magnitude 19, while by means of photography they go down to magnitude 21. It requires 100 million stars of magnitude 21 to give as much light as a single star of the first magnitude – for instance, Altair or Aldebaran.

Remember that the word *magnitude* as used here does not refer in any way to the size of the stars, only to their brightness.

Return now to Polaris. It is at the end of the long tail of Ursa Minor (the Little Bear). When the stars are in the position shown in Fig. 101 the bear is standing on its head. Polaris is of the second magnitude. The two stars of this constellation next to Polaris in brightness are in the right shoulder and foreleg of the animal, and are known as the "Guardians of the Pole." The brighter one is of the second, the other of the third magnitude.

Cassiopeia and her Neighbours

On the other side of the celestial pole from Ursa Major is the constellation Cassiopeia. Five of its chief stars form a sprawling M, or a W turned upside down. Three of the neighbouring constellations are named Cepheus, Andromeda, and Perseus. These names are all well known in the old mythology.

According to the Greek legend, Andromeda was the daughter of Cepheus and Cassiopeia, who were the King and Queen of Ethiopia. Cassiopeia having boasted herself to be equal in beauty to the sea-nymphs who constantly waited upon Neptune, the god of the sea, that august person became angry, and to punish the King and Queen he sent a great flood upon the land, and also let loose a great sea-monster which destroyed the people and the animals of the kingdom. Neptune let it be known that no relief would be given until the King gave up his daughter to be torn to pieces by the monster. So the luckless lady was taken to the shore of the sea, chained to a rock, and left there to her fate. Shortly after this a fine young man named Perseus, who was returning home from a wonderful victory over the fabled Gorgon, saw Andromeda in her sorry plight, slew the monster, set her free, and then married her[1]. The ancient people imagined that they could trace the shapes of these fabulous persons in the stars, but modern observers fail to see anything of the kind.

Between Ursa Major and Ursa Minor is Draco (the Dragon). The tip of its tail is not far from the line running from the Pointers up to Polaris. Stretching from there the long body is traced out by a series of comparatively faint stars. As we follow along its course we pass over a coil in the body, and at last come to the head, which is not far from the constellation Hercules. The head is marked by four stars forming an irregular four-sided figure. The brightest of these four stars (which is the one farthest from Polaris) is denoted by the Greek letter γ (*gamma*). As the sky turns about its axis every day this star comes almost directly over London, and some very important observations were made on it by James Bradley, who was Astronomer Royal from 1742 to 1762, the third to hold the office.

Another notable star in Draco is *a*, or Thuban, which is midway between Mizar and the Guardians. Some 4700 years ago it was our Pole Star. At that time Polaris was much farther from the celestial pole than it is now.

With a map like this one it is easy to pass from these constellations to the others and thus to identify them all. Great pleasure may be had in tracing them out.

[1] The story is well told in *The Heroes*, by Charles Kingsley.

The Equatorial Stars – The Stars of Winter

Let us now face the south and study the stars contained in a wide belt along the celestial equator. This belt is long, since it runs completely round the celestial sphere, and we shall consider it in four parts.

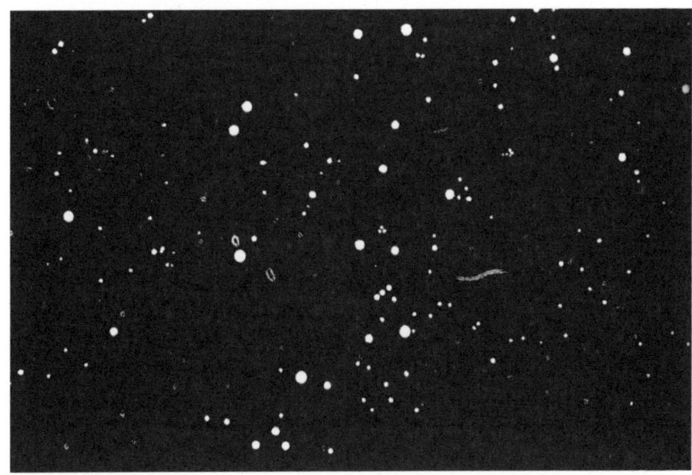

FIG. 103. THE WINTER STARS

We shall begin with the winter stars, as they are the finest of all. In the map in Fig. 103 the stars of winter are shown, while in Fig. 104 is a key to the names. The straight line across the latter is the celestial equator, while the curved broken line is the ecliptic, which, you remember, is the path in the sky which the sun appears to follow.

The sun moves (from right to left) along that portion of the ecliptic which is shown here during the interval from April 7 to September 7; and midway between these two dates – that is, on June 21 – it is at that point which is highest above the celestial equator.

Think of a plane which passes through the celestial pole, the point directly overhead (the zenith) and yourself, the observer.

It will also pass through the N. and S. points of the horizon. This is your *meridian plane*. It cuts the celestial sphere in a great circle, and this circle is the observer's meridian. When the sun, or any other heavenly body, is on this circle, it is said to be 'on the meridian.'

At the top of the key-map (Fig. 104) are the names of the months. The stars under the name of a month are on the meridian at about 9 P.M. during that month.

The most prominent winter constellation is Orion, whom the ancients represented as a mighty warrior. We shall see his picture a little later (Fig. 117). The three bright stars almost in a straight line and about 1½° apart are in his belt. They are of the second magnitude. Then 8° north of the belt is a fine red star named Betelgeuse. It is of the first magnitude. Ten degrees south of the belt is a splendid blue-white star which bears the name of Rigel. It also is of the first magnitude. The constellation of Orion is the finest in the entire sky. It is the only one which contains two first magnitude stars.

The Great Dog, the Bull, and the Twins

Next continue along the belt to the left about 20° and you come to Sirius, the Dog Star, in the constellation of Canis Major (the Great Dog).

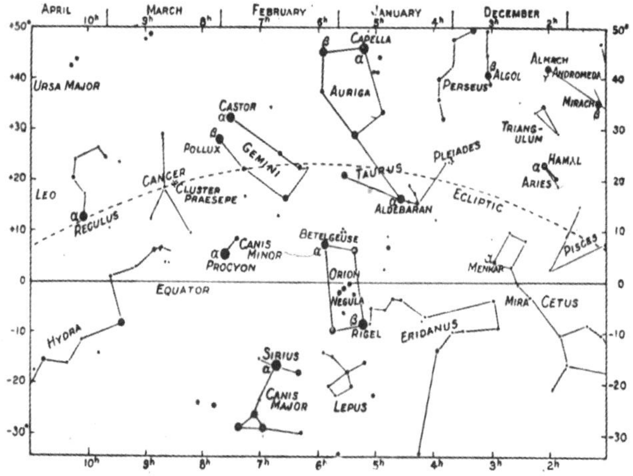

FIG. 104. KEY TO THE WINTER CONSTELLATIONS

FIG. 105. THE SPRING STARS

It is the brightest star in the heavens. Then, following the belt to the right about 20°, you come to Aldebaran, a red star of the first magnitude in the constellation Taurus (the Bull).

Orion is just on one side of the Milky Way. On the otherside, higher than (that is, north of) and to the left (that is, east) of Orion is the constellation Gemini (the Twins). The two brightest stars in it are Castor and Pollux. The former is of the second magnitude, the latter of the first. Travelling from Castor through Pollux and swerving a little toward Orion, you reach Procyon, a bright, first – magnitude star in Canis Minor (the Little Dog). Some distance to the left of the line joining Pollux and Procyon, and forming a great triangle with these two stars, is Regulus – in Leo (the Lion). Regulus is also a first-magnitude white star, and is almost exactly on the ecliptic. Regulus and five more stars in Leo form the Celestial Sickle.

The other stars on the map can easily be located in the sky by working from those which we have spoken of.

The Stars of Spring

In the next map (Fig. 105) we have the stars of spring. Their names are given on the key-map (Fig. 106).

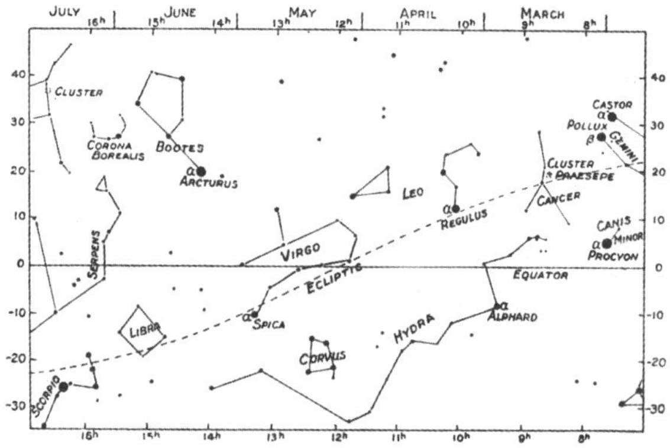

FIG. 106. KEY TO THE SPRING CONSTELLATIONS

In March the constellation Cancer (the Crab) is in a good position to be seen, but it has no bright stars in it. About half-way between Pollux and Regulus is a cluster of stars which is named Praesepe (the Beehive). It is interesting to try to find it. On a dark night it looks like a small, nebulous cloud, but an opera-glass or a small telescope reveals a host of separate faint stars. It is a coarse cluster, and we shall see a picture of it later (Fig. 135).

In April and May Virgo (the Virgin) is easily located. It contains one bright white star named Spica, which is near the ecliptic and about 10° south of the equator. This star Spica is a very great sun. According to the best measurements, its distance is 360 light-years – that is, light takes 360 years to come from it to the earth. We see it as it was 360 years ago. We receive from our sun many million times as much light as we do from Spica, but that is simply because the latter is so much farther off. Indeed, if it could be placed at the same distance as our sun it would give us 3000 times as much light.

But the most notable of the stars which come to us in the spring is Arcturus, the chief attraction in the constellation Boötes (the Herdsman). This is one of the first stars to which a name was given, and it has been an object of interest and admiration from the earliest times. It is mentioned by the Greek poet Hesiod, who lived 800 years before Christ.

FIG. 107. THE SUMMER STARS

On March 1 Arcturus rises in the east at about 8 P.M., and as it rises four minutes earlier each evening at the end of the month it comes up as the sun goes down. At that time it is visible all night long, and is a prominent object even when the moon is full.

Arcturus is 41 light-years from the earth. It is called a *fixed* star, but careful observation of its position shows that it is moving on the celestial sphere and that its speed is 85 miles a second. Yet, on account of its great distance, it will require 800 years for it to be displaced the width of the moon. In very recent years its diameter has been measured. It is 25,000,000 miles – thirty times our sun's diameter – and it gives out 105 times as much light as the sun.

Perhaps you will recall some of these remarkable facts when next you gaze into the face of Arcturus.

The Stars of Summer

We now come to the stars of summer (Figs. 107, 108). One of the constellations seen at this season is Lyra (the Lyre or Harp), and though small it has some very interesting features. This is the lyre on which, according to ancient myth, Orpheus played such sweet music that the wild beasts became tame, and the rivers, trees, and rocks moved toward

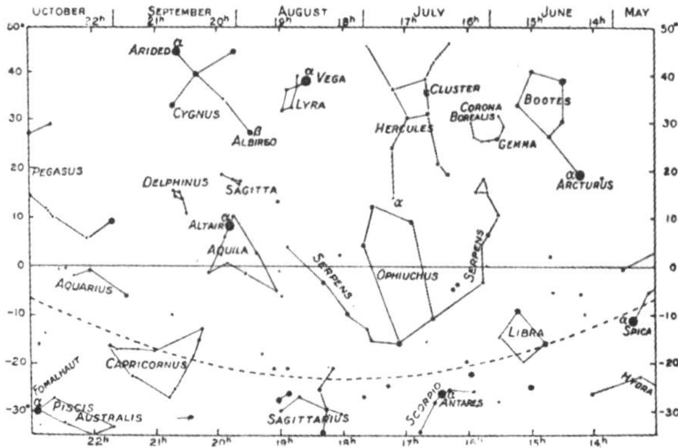

FIG. 108. KEY TO THE SUMMER CONSTELLATIONS

him. By it also he so softened the stony hearts of the keepers of the lower world that they permitted his dead wife Eurydice to return to earth with him. After his death his lyre was placed in the sky.

The gem of the constellation is Vega, a splendid blue-white star. In the middle northern latitudes it is almost overhead in the early evening at the beginning of August. Of all the stars visible in the northern hemisphere Sirius is the brightest and Vega is next, though it is followed closely by Arcturus and Capella. Vega and five other stars form a little equilateral triangle and a parallelogram, which are easily recognized. Between the two stars of the parallelogram farthest from Vega the Ring Nebula (Fig. 125) is located, but it cannot be seen with the naked eye.

It has been proved that our whole solar system is moving through space at the rate of 12 miles per second, and that its course is directed toward the constellation Lyra.

East of Lyra is Cygnus (the Swan). It is sometimes known as the Northern Cross since five prominent stars in it trace out a distinct cross. The Northern Cross is in the midst of a bright portion of the Milky Way. The star at the head of the cross is called Arided (or Deneb). It is one of the most remote of the bright stars, being probably over 600 light-years distant; and it gives out, it is believed, 10,000 times as much light as our sun. The star at the foot of the cross is named Albireo.

FIG. 109. THE AUTUMN STARS

Between Lytra and Boötes are Hercules and Corona (the Crown). In Hercules four stars, not very bright, form a Flowerpot or a Keystone. In one side of the 'pot' is a faint star which you can just see with the naked eye. In a small telescope it looks hazy, and you hardly know what to think of it. In a large telescope it looks like a magnificent cluster of jewels. There are photographs of it elsewhere in the book (Fig. 135 and frontispiece). Arcturus is still a fine object in the sky.

Down near the equator is Altair in Aquila (the Eagle), and much farther south is the red star Antares in Scorpio (the Scorpion). This star is believed to be four hundred million miles in diameter!

The Stars of Autumn

Lastly let us look at the autumn sky (Figs. 109, 110). Lyra and Cygnus can still be seen, somewhat west of the zenith – the point in the sky directly overhead. High in the southern sky is Pegasus, the winged horse ridden by Perseus when he went out to fight the fierce monster which intended to destroy Andromeda. There are four bright stars which form the Great Square of Pegasus. Actually one of these is in the adjoining constellation of Andromeda. It bears the name Alpheratz. The two other chief stars of Andromeda are named Mirach and Almaach. These three are nearly in a straight line. The Great Square

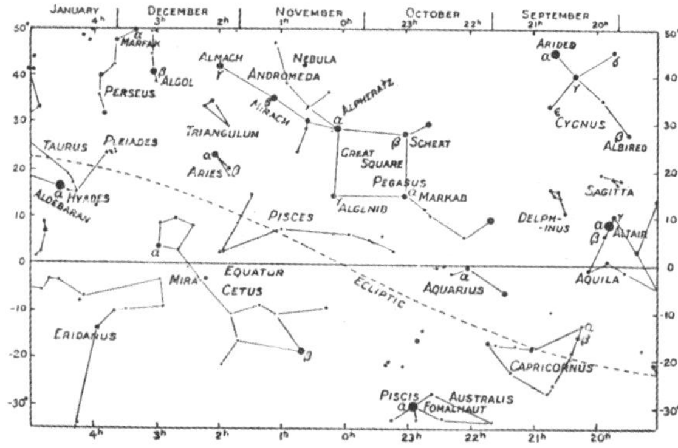

FIG. 110. KEY TO THE AUTUMN CONSTELLATIONS

resembles the body of a kite with the three stars of Andromeda as tail. Indeed, the tail may be extended to Marfak, the brightest star in Perseus. About 10° west of Mirach (near the letter E in "Andromeda" in Fig. 100) is a wonderful object known as the Great Nebula in Andromeda. We shall have a photograph of it later (Fig. 121).

To the east of Andromeda is Perseus, which contains a remarkable star named Algol, an Arabic word for "the Demon." This star is usually as bright as the Pole Star, but every sixty-nine hours it loses three – fourths of its light, and in a few hours recovers it again. It does this with perfect regularity. In this star are two bodies, a bright one and a dark one, revolving about each other. During every revolution the dark one comes in front of the bright one and shuts off three – fourths of its light; but as soon as the partial eclipse is over the bright one shines as usual.

About 45° southward from Algol is a star in Cetus named Mira. Now *mira* is the Latin word for ' wonderful,' and a wonderful star it is. Usually it is too faint to be seen with the naked eye, but every eleven months it becomes almost as bright as the Pole Star. No satisfactory explanation of this star's behaviour has been given.

Still farther south is seen on October and November evenings a white star in Piscis Australis (the Southern Fish) which bears the name Fomalhaut. It is so far south, being 30° below the celestial equator, that it is above the horizon only a few hours. There are no bright stars near it, and so it is easy to identify.

CHAPTER X

THE NUMBER OF THE STARS; THEIR DISTANCES – THE NEBULÆ

The Number of the Stars

HEN, out in the country and away from the city lights, you stand under the open sky on a perfectly clear moonless night, the stars seem countless in number. But are they really countless? One way to test this is to choose some definite area of the sky, such as the Bowl of the Dipper; or the Square of Pegasus, and actually count the number of stars in it. You will find there are comparatively few.

The stars visible to the naked eye have been counted by several astronomers, and the number is about 6000. Now one can see only half of the celestial sphere at any time, and hence about half of this number, or 3000, will be above the horizon at any moment. Also, the haze in the atmosphere near the horizon always prevents the faint stars there from being seen, and so it is probable that not more than 2000 can be seen at one time by the unaided eye. In a city, where the lights of the streets and houses add to our difficulties, it is doubtful if as many as 1500 can be seen at once. That is far from countless!

A simple opera-glass enables one to see at least 100,000, while our largest telescopes show probably more than 100 million, and long-exposure photographs reveal many more. There is certainly an enormous number of stars, but not an *infinite* number – that is, a number which cannot be calculated.

Stars seen with the Naked Eye and with a Telescope

About seventy-five years ago a German astronomer, Argelander, made charts of all the stars north of the celestial equator which he could see with a telescope 7 cm. (approximately 2¾ in.) in diameter. There were 324,198 in all. In Fig. 111 is shown a portion of one of his charts. It covers an area 8° square, and includes 1442 stars. The two

FIG. 111. STARS SEEN WITH A SMALL TELESCOPE

This is a portion of one of the famous star-charts made by Argelander, a German astronomer. It shows all the stars visible in this part of the sky with a 2 3/4 inch telescope. It comprises the northerly portion of Orion, the three bright stars in the giant's belt being semi at the bottom. This portion is 8° square, and hence the area is 64 square degrees[1]. It contains 1442 stars, on average of 22½ to a square degree.

small squares in Fig. 112 show the stars which can be seen in the same area with the naked eye. In the right-hand one are those which can be seen on a clear, moonless night if you are out in the country twenty-four stars. In the left-hand one are those which can be seen out in the country if the moon is full, or in the city with its bright lights in the street and houses – only eight stars! In Fig. 113 is shown a photograph of this same part of the sky. The white frame marks off the area included in Fig. 111, and it is simply impossible to count the stars which are in it. Notice that the sky all about here is filled with a nebulous haze.

The stars in these pictures are those in the Belt of Orion and in the space just north of it. The three stars in the Belt are at the bottom. It is interesting to pick out those stars in Figs. 111 and 113 which can be seen with the unaided eye.

[1]A square area of the sky with sides 1° in length is it square degree.

In some parts of the sky the stars are much more numerous than in others. This is especially marked in the Milky Way, the faint band across the sky which has been referred to already and which is familiar to all.

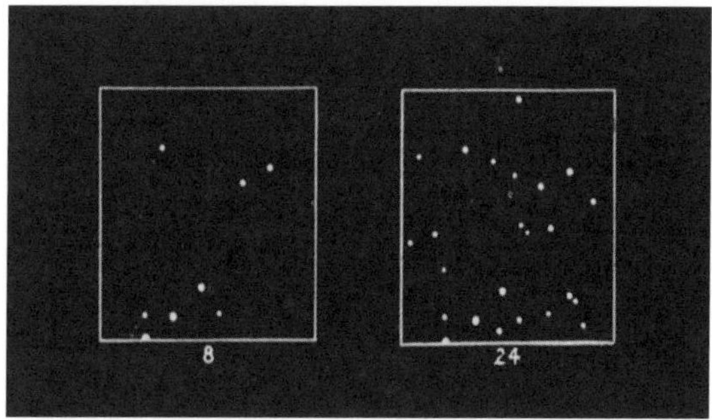

FIG. 112. THE SAME AREA OF THE SKY SEEN WITH THE NAKED EYE
These two little squares show the stars which can be seen with the naked eye in the same space as covered by Fig. 111. Right: those seen in the country on a clear, moonless night – twenty-four stars. Left: those seen in the city with its artificial lights or in the country at full moon – eight stars.

A Picture with Half a Million Stars in it

In Fig. 114 is shown a portion of the Milky Way in the constellation Sagittarius, which is south of the equator. The original photographic plate was 14 by 17 inches in size, and on it some 2,000,000 stars were recorded; on that part of it in Fig. 114 there are 500,000 stars.

In Fig. 115 is another portion of the Milky Way. The stars at the centre seem packed together so closely that they form a white cloud. Note the two bright streaks – one at the top, the other at the bottom. They were produced by two shooting stars which happened to come along while the astronomer was taking his long-exposure photograph.

Some of the Constellations – The Nebulæ

We have already learned how to pick out the constellations. Let uts look at a few of them more closely.

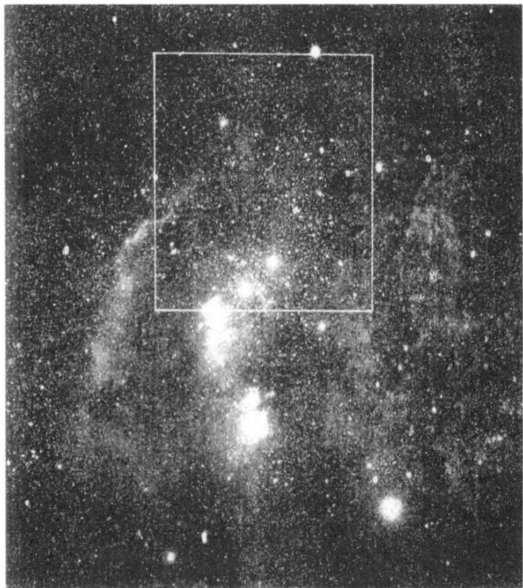

FIG. 113. A PHOTOGRAPH OF THE STARS SHOWN IN FIGS. 111, 112
This is a photograph of a part of the constellation of Orion, taken with a lens designed by Ross, with a diameter of 3 inches and focal length of 21 inches. The exposure was five hours. The area within the white frame is the same as that in the previous cases. It is simply impossible to count the stars – surely there are 200,000 at least! Notice also the nebulous matter which seems to fill much of this constellation. The bright star half an inch from the bottom and an inch from the right-hand edge is Rigel.
Photograph by F. E. Ross, Yerkes Observatory

The Great Bear

One of the best known is Ursa Major (the Great Bear). In Fig. 116 the stars in this constellation visible to the naked eye, and also the animal as the ancient astronomers pictured it up in the sky, are shown. Anyone who can see the form of a bear in these stars must have a lively imagination.

You notice that the Dipper, or Plough, is only a small part of the constellation, the handle of the Dipper being the tail of the Bear. Why the Bear is represented with it long tail it is impossible to say.

FIG. 114. A PHOTOGRAPH SHOWING HALF A MILLION STARS

The portion of the Milky Way covered by this photograph is in the constellation of Sagittarius. The size of the original negative was 14 by 17 inches. It covered an area of 65 square degrees, and each of these contained about 45,000 stars, or a total of over 2,000,000 on the entire plate. The present picture is a portion of the original, covers about 22 square degrees, and contains over 500,000 stars. *Harvard Observatory photograph*

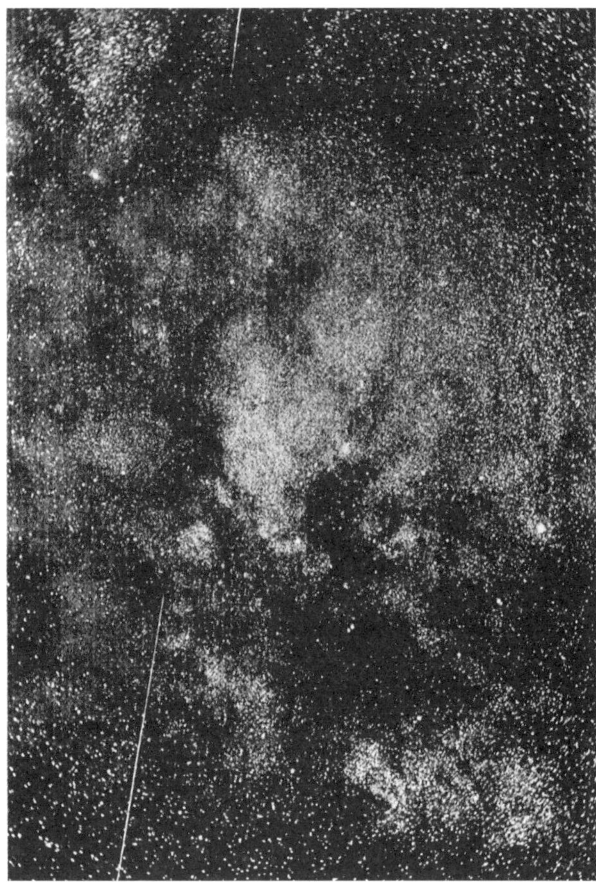

FIG. 115. STAR CLOUDS IN THE MILKY WAY, WITH TWO METEORS

This dense star cloud is in the southern part of the constellation Aquila. The photograph was taken on Mount Wilson, California, with the 6-inch Bruce telescope of the Yerkes Observatory. The exposure was 2 hr. 40 min. Note the two bright trails made by meteors which came along during the long exposure and left their marks on the photographic plate.

Photograph by Barnard

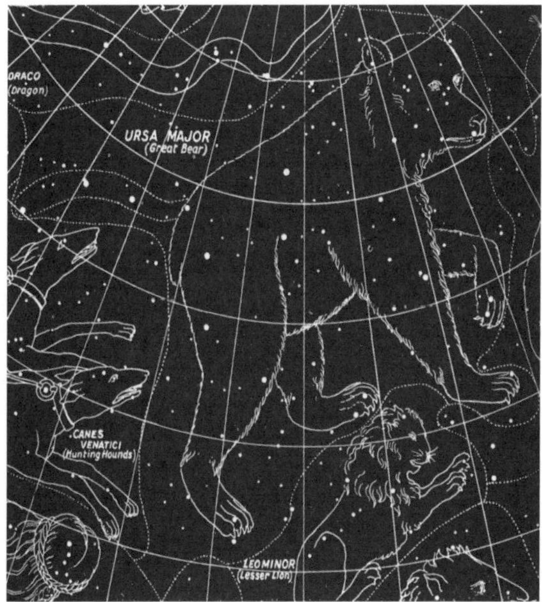

FIG. 116. THE CONSTELLATION URSA MAJOR (THE GREAT BEAR)

The animal is drawn as the ancient astronomers imagined they could see it. The Plough, or Dipper, is only a small part of the constellation. A scattered group of small stars is at the head of the bear, while pairs of stars mark out three of its paws. The two outer stars of the 'bowl' are the pointers. The upper one (as seen in the picture) is called Dubhe, the other Merak. The star at the bend of the handle is Mizar, and the little star very near it is Alcor, "the rider on his horse".

Orion

In the case of Orion (Fig. 117) the great giant seems to fit in among the stars somewhat better. Notice the three bright stars in the belt from which hangs his sword. There are many fine stars in this constellation. The brightest is Rigel, which is pearly white. It is in the giant's left foot. It is 540 light-years distant, and gives out, it is estimated, 17,000 times as much light as our sun. Betelgeuse is next in brightness, and is orange red in colour. It is in his right shoulder.

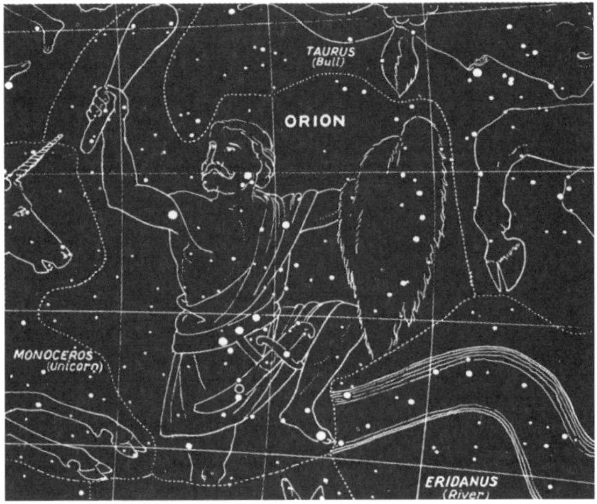

FIG. 117. THE CONSTELLATION ORION

Orion is the finest constellation in the sky. The giant stands with his club ready to drive back to the Bull. In his right shoulder is the giant red star Betelgeuse, 240,000,000 miles in diameter. In his left hand he holds up the lion-skin, traced out by a curving line of stars. In his left shoulder is the white star Bellatrix, and in his left foot is the blue-white star Rigel, a great sun 17,000 times as luminous as our sun. Note the three stars in the Belt - in a straight line and about 1½° apart. They are sometimes known as the "Ell and Yard".

In recent years astronomers have succeeded in measuring the diameter of Betelgeuse, and the result is astonishing. The diameter is 240 million miles! If the whole solar system could be carried away and placed with the sun at the centre of Betelgeuse the earth would be 30 million miles within its outer surface and Mars would move along the surface of the star! But remember Betelgeuse is composed of thin gas.

The white star in Orion's left shoulder is Bellatrix. Three stars are in his left jaw, while some ten stars in a curved line trace out the lion's mane which he holds as a shield on his left arm.

Note the little circle on the sword. This shows the position of a remarkable object. With the naked eye it looks like a star, in a field-glass it looks like a hazy spot, but in a large telescope or in a good photograph it is simply wonderful. In Fig. 118 is a picture of it.

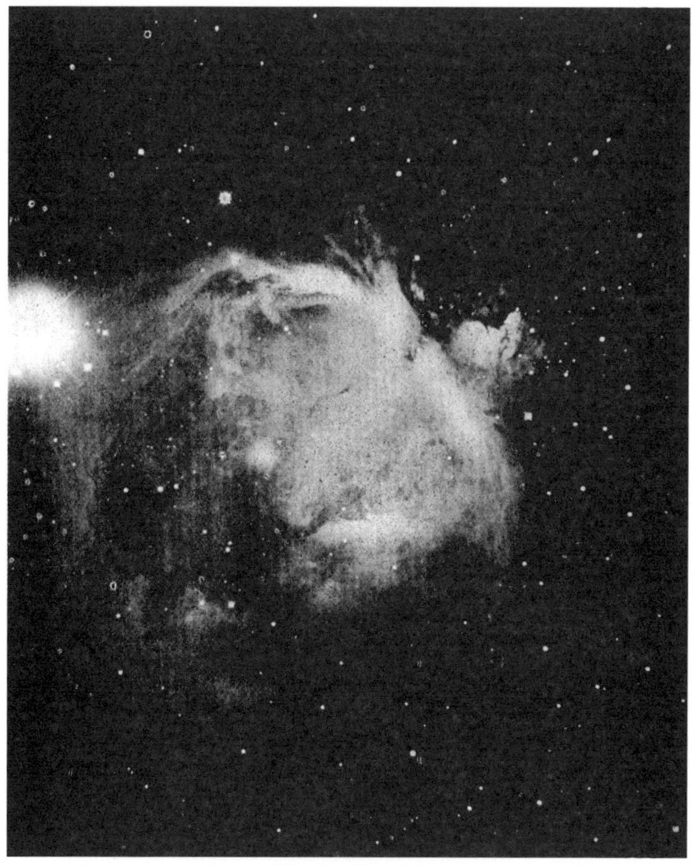

FIG. 118. THE GREAT NEBULA IN ORION

This wonderful object is in the sword hanging from Orion's belt, at the place marked by a little circle in Fig.177. The dark markings shown in this fine photograph are probably clouds of dust obscuring the bright surface behind them. The nebula is believed to be at least 600 light-years distant. This photograph was taken on November 19, 1920, with an exposure of three hours.

Photograph with 100-inch telescope, Mount Wilson

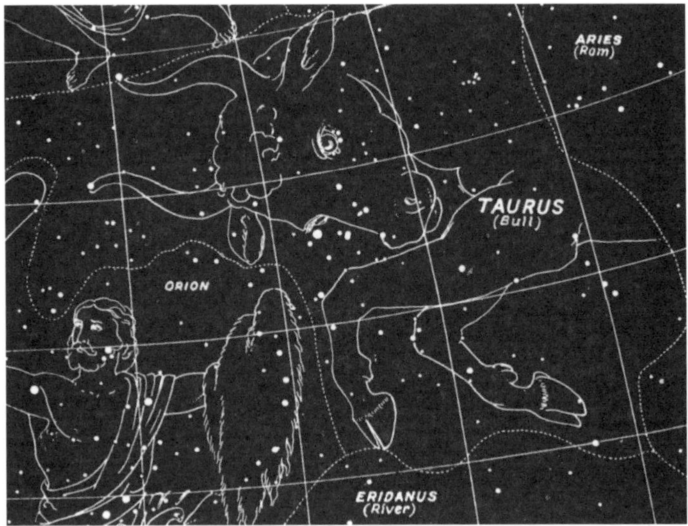

FIG. 119. THE CONSTELLATION TAURUS (THE BULL)
The fierce beast is charging down upon Orion. In its right eye is the blazing red star Aldebaran. This mighty sun is fifty-seven light-years away, is recoding from our system at the ratio of 54 miles per second, and it has a diameter of 32 million miles. It is the chief star in the V-shaped group called the Hyades.

This object is called the Great Nebula in Orion. *Nebula* is a Latin word for 'cloud'. This nebula has no regular shape, but spreads out in all directions in great curved sheets and irregular masses. What is it made of? There is much hydrogen gas in it, and also much gas of a sort not yet fully understood by our earthly chemists.

Taurus

You remember Orion had a big club in his right hand. He has it drawn back ready to defend himself from the great Bull which threatens to charge him (Fig. 119). Notice the large star just at the Bull's right eye. That is Aldebaran. On the Bull's neck is a little group of six stars. It bears the name of the Pleiades. These stars have been observed in many widely separated countries for thousands of years. From the earliest times they were used to regulate the calendar and to determine the times of religious feasts by the natives of Peru, India, Australia, Egypt, and many other countries.

Though the individual stars are faint and not easily seen, the group is conspicuous and at once attracts the attention. It is well seen in the autumn, as it rises in the east, or in the spring, as it sets in the west.

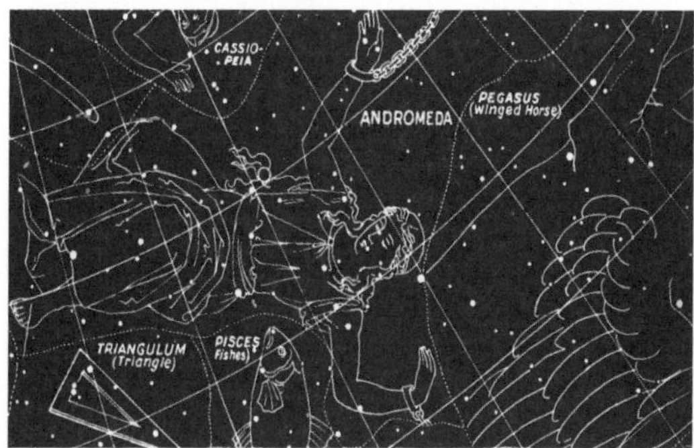

FIG. 120. THE CONSTELLATION ANDROMEDA

This constellation is recognized by its three bright stars almost in a straight line. α (or Alperatz) is at her head, β (Mirach) is at the left side of her waist, and γ (Almaach) is at the edge of her skirt, near the left foot. The circleat the right side of the waist shows the position of the Great Nebula. Just to the north is Cassiopeia, the motlier of Andromeda. Also note the Triangle just below.

We shall have some photographs of it later (Figs. 133, 134).

Tennyson has, in *Locksley Hall*, a very striking and true description of this famous little group:

> Many a night I saw the Pleiads, rising thro' the mellow shade,
> Glitter like a swarm of fire-flies tangled in a silver braid.

Andromeda

Here is poor Andromeda chained to the cruel rocks (Fig. 120). The bright star near her left ear is Alpheratz, the one at the left side of her girdle is Mirach, while Almaach is at the edge of her skirt.

FIG. 121. THE GREAT NEBULA IN ANDROMEDA

This may clearly be seen to be a nebula with the naked eye, and it is a fine object in the telescope. It is 3° across – six times the diameter of the moon. Its distance is estimated to be 850,000 light-years, and its diameter 45,000 light-years. It is perhaps as large as our entire Milky Way, and may be an "island universe" by itself. The separate stars apparently scattered over the nebula are in front of it. This nebula is a spiral, seen obliquely. It was photographed with the 2-foot reflector, with an exposure of four hours.

Photograph by Ritchey, Yerkes Observatory

FIG. 122. THE SPIRAL NEBULA IN THE TRIANGLE

This is a faint object, and its remarkable shape and beauty can be revealed only in a photograph. It is a spiral seen full face. The nebulous patches in the arms are inter-mingled with stars; indeed, it is believed that the nebulous matter is condensing to form stars. It is about as distant as the Andromeda Nebula, and its diameter is about 15,000 light-years. The photograph was taken with the 60-inch reflector, with an exposure of eight-and-a-half hours.

Photograph by Ritchey, Mount Wilson Observatory

About 10° above Mirach, near the right side of her girdle, there is a little circle. This marks the position of another of those strange nebulae. In Fig. 121 is a photograph of it. This is the Great Nebula in Andromeda. It is the only one which can be recognized as a nebula by the naked eye.

This famous object is well placed for observation during the evening, in the summer in the east and in the winter almost overhead. In a field-glass it looks like an oval patch of bright cloud, and its true beauty and magnificence is revealed only by photography. This is a fine picture of it.

Is this nebula also composed of gas? The spectroscope says no. Indeed, there is reason to believe that it is simply a multitude of stars, and that it is probably as large as our Milky Way, but so far away that it looks like a little cloud in the sky. It is approaching our system at the rate of 200 miles per second.

The Triangle and its Nebula

Just next to Andromeda is a small constellation named the Triangle, which contains another famous nebula. This nebula is shown in Fig. 122, and is a lovely spiral. Here you can see the central nucleus and the two arms running out from opposite sides of it. It looks much like a Catherine-wheel spinning round and throwing oil sparks. Surely this nebula is rotating! And how long will it require to turn completely round? About 160,000 years!

Other Spiral Nebulae

Here is another fine spiral (Fig. 123). It is in the constellation Hunting Hounds, and is not far from the end of the handle of the Dipper. As before, there are two arms starting out from the great central nucleus and winding themselves about it.

This nebula cannot be seen with the naked eye, and through even a good telescope its form cannot well be made out. But what a beautiful object the photograph shows it to be! This one also has been examined for motion, and its period of rotation has been given as 45,000 years. It is moving as though to wind up the arms – the top is moving to the right – the reader's right as he looks at the illustration.

We are looking directly at these last two spirals. The nebula in Andromeda is seen with face turned somewhat from us, but Fig. 124 shows a spiral with edge turned toward us. In this position it looks quite thin. The nucleus, we see, is nearly spherical, and the dark streak down the middle is the very edge of the nebula. Slipher, at the Lowell Observatory, has shown that this nebula is moving away from our system at the rate of 625 miles per second.

FIG. 123. THE "WHIRLPOOL" NEBULA IN THE HUNTING HOUNDS

A famous spiral looked at directly. Note the bright central nucleus with the two arms running out from opposite sides of it. These are composed of nebulous matter and stars. The spiral nature of this object was first detected in 1845 by the Earl of Rosse in his famous 6-foot reflector; but not until photographs of it were taken by modern telescopes did we find our the details of its structure.

Photograph by Humason, 100-inch telescope, Mount Wilson Observatory.

FIG. 124. A SPIRAL NEBULA SEEN EDGE ON

This strange object is in the inconspicuous constellation known as Berenice's Hair, between Boötes and Leo. The photograph shows it to be a spiral nebula viewed edgewise. Note the bright nucleus which is evidently shaped like a ball. The dark band through it is without doubt the edge of the nebula, or one of the arms in front of the bright nucleus. The photograph was taken with the 60-inch reflector and the exposure was five hours.

Photograph by Ritchey, Mount Wilson Observatory

There are thousands of these spiral nebulae in the sky all at immense distances. They seem to be the common types of nebula. The thoughtful mind cannot but wonder how they were made and what they will become.

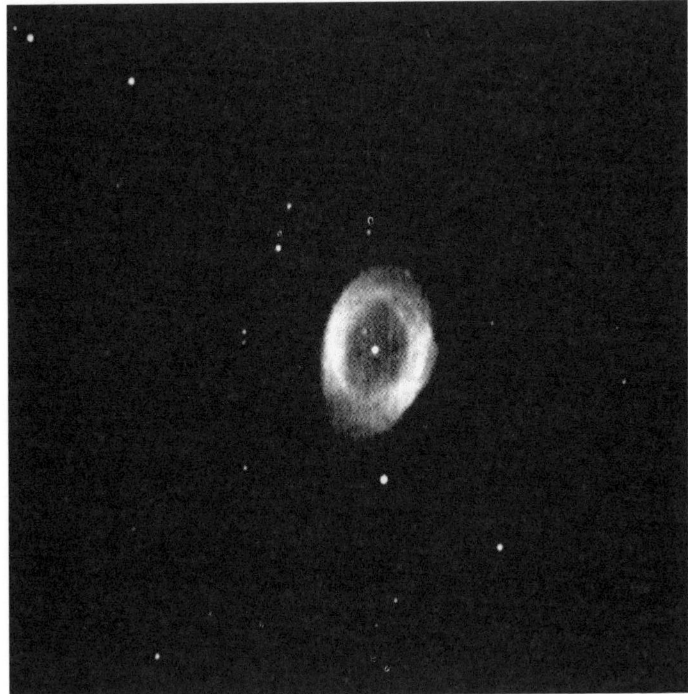

FIG. 125. THE RING NEBULA IN LYRA

This nebula is between the stars β and γ of Lyra, and can be seen with a small telescope. The central star which is so prominent in the photograph is extremely faint visually, and can hardly be seen in a 2-foot telescope. The average distance of these bodies is probably about 400 light-years. The nebula shown in this photograph would almost certainly cover an area greater than the entire solar system. The photograph was taken with the 6-foot reflector of the Dominion Astrophysical Observatory, Victoria, British Columbia, and the exposure was thirty minutes.

Photography by Plaskett, Victoria, British Columbia

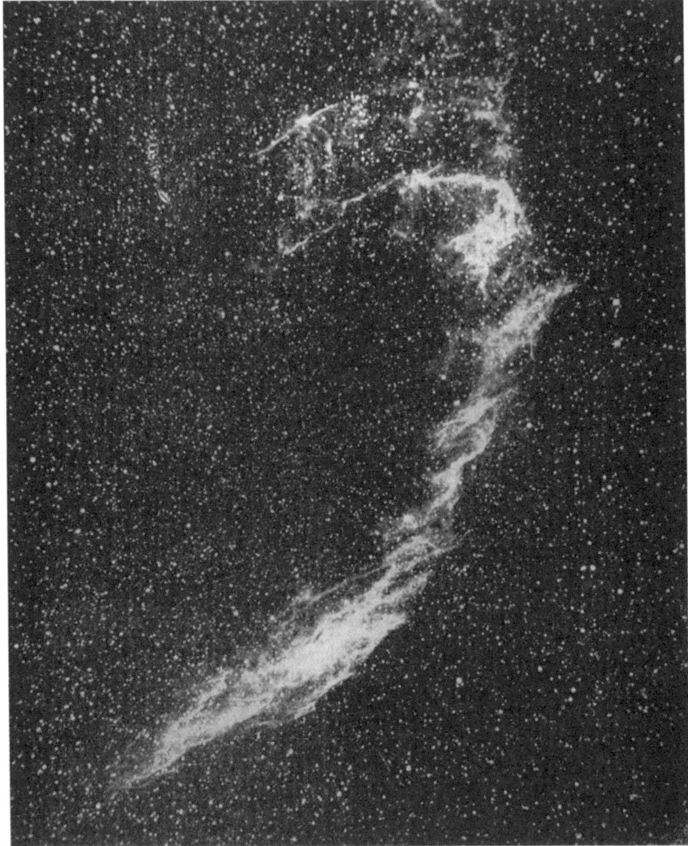

FIG. 126. THE "NETWORK" NEBULA IN CYGNUS (THE SWAN)

This irregular filamentary nebula is in the midst of the Milky Way. It is probably several hundred light-years away, but not so distant as the spirals. Not far from this nebula there is another of the filamentary type, and photographs reveal the existence of much extremely faint nebulous matter in the space about them. This photograph was taken with the 2-foot reflector, with an exposure of three hours.

Photograph by Ritchey, Yerkes Observatory

Other Kinds of Nebulae

There is another style of nebula found in Lyra (the Harp). A small telescope shows its shape very well, but a photograph reveals it much better (Fig. 125). It is shaped like a ring. Other ring nebulæ have been discovered, but none as fine as this one.

There are many other strange nebulæ, but at present we shall look at only one more (Fig. 126). It is named the "Network" Nebula, and is found in Cygnus (the Swan). It looks like filmy lace blown about by the wind, but you must remember that where this nebula is located there are no air-currents such as we have on the earth.

The Distance of the Stars

Up to the present we have said little about the distance from the earth of the stars and nebulæ. It is only in recent years that we have been able to find out just how far they are from us. Measuring their distances is about the hardest task the astronomer has to undertake. It requires the utmost skill and patience and care – all because the stars are so far away.

FIG. 127. "THE WINGS OF THE MORNING"
The light from the rising sun streaming through the openings between the clouds at the eastern horizon produces the fan-shaped system of sunbeams shown in this beautiful picture. They are undoubtedly "the wings of the morning" referred to by the Psalmist.
Photograph by Ballance, adapted

We have learned that the sun is 93 million miles from the earth, but that means little to the ordinary person; and when we find out that the nearest of the stars is 272,000 times as far away as the sun, we are simply lost in the figures.

Let us try another way to appreciate what this means.

"The Wings of the Morning"

In the Bible there are many beautiful poetic phrases, or figures of speech, and one of the most striking is to be found in Psalm cxxxix: "If I take the wings of the morning, and dwell in the uttermost parts of the sea; even there shall Thy hand lead me, and Thy right hand shall hold me."

Now what is meant by "the wings of the morning"? The photograph in Fig. 127 explains this very well. Just as the sun rises above the eastern horizon its beams shine upon the clouds and pass through the spaces

FIG. 128. FLYING INTO THE SUNSET
This is a war-time photograph taken at Camp Borden, Ontario. It illustrates well the familiar phenomenon known as ' the sun drawing water' which is produced by the sun's rays passing through openings in the clouds, and is similar to "the wings of the morning."

between them, painting them in glowing colours and producing the straight streamers which we see here. These are "the wings of the morning." You no doubt have often seen something similar in the afternoon in the western sky. The phenomenon is then often described as 'the sun drawing water' (Fig. 128).

To "take the wings of the morning" then must mean to receive the power of travelling through space with the speed of light. And how great is that? Though it is extremely high, this speed has been measured accurately. It is 186,284 miles per second, or about eleven million miles per minute.

We think sound travels very quickly, but it takes five seconds to go a single mile, and to go round the earth would take thirty six hours. Light, however, can encircle our globe seven times in a single second.

Suppose then we are supplied with "the wings of the morning" and, thus equipped, make an extended trip through space, travelling with the speed of light and visiting some of the wonders of the universe.

Let us start at the sun, the centre of our system. In three and a third minutes we should reach Mercury, in six minutes Venus, in eight and a third minutes the earth, and in thirteen minutes Mars. Then we move on to Jupiter, Saturn, and Uranus, and after four hours from starting out we reach Neptune.

We are now on the outskirts of the solar system, and we look round us in order to choose a suitable star to visit.

We might suppose Sirius to be the nearest, since it is the brightest; but, perhaps remembering that an astronomer on the earth had stated that a star in the southern skies named Alpha of the Centaur was the nearest star to our system, we start for it.

We go forward a full day; it seems no brighter, it looks just as far off. We go on for a week, a month – no noticeable change. That star must be a tremendous distance away! We continue a year, always rushing forward at the rate of eleven million miles every minute. We are certainly getting closer, since it is almost twice as bright as it was at first, but it is not yet as bright as Sirius is.

So we go on for two years, three years, four years; it is much brighter now, but still only a star. But at the end of four months more it is very much brighter, and at last we are close enough to examine it carefully.

And what do we find? It is a great big, brilliant sun like our own!

So it is with the rest of the stars. They are all suns – some larger, some smaller than our sun; and doubtless some of them have planets revolving about them, though of this we have no certain knowledge. Centaur with the speed of light, we say it is four and a third light-years away.

A Cobweb to a Star

The late John A. Brashear, of Pittsburgh, the famous maker of lenses and mirrors for telescopes, used to tell an interesting story.

In the eyepiece of some telescopes it is necessary to insert a very fine thread, or wire, to assist in making measurements, and for many years spider-web has been used for this purpose. The instrument-maker does not use the rather coarse webs which we see on trees or grass covered with dew in the morning, but the fine fibre which the mother spider wraps round the little cocoon to strengthen it and to protect the young within. This is extremely slender and delicate. To handle it demands care and skill.

One day a workman who had been using some of the web weighed a measured length of it on a delicate balance, and from this he computed how far a pound of it would reach. He found that that weight of web would be 25,000 miles in length, and that it would encircle the earth, while ten pounds would stretch beyond the moon.

Dr Brashear then calculated how much would be needed to reach to Alpha of the Centaur. How much do you think? It would take 500,000 tons! To ship it by railway would require a train 150 miles long, drawn by 500 powerful locomotives.

And that is the distance to the *nearest* of the fixed stars – to our next neighbour among the suns of space. Sirius, the brightest of them, is over eight light-years away; Vega is twenty-six, while the Pole Star is 466. And there are many beyond that!

You see, then, that the sun and its system occupies a very small portion of space; while in all directions, at inconceivable distances, are the stars and nebulae.

It is a grand universe – magnificent, stupendous!

CHAPTER XI

DARK MARKINGS – CLUSTERS – THE NATURE OF THE STARS

Dark Holes and Markings in the Milky Way

HERE are other wonderful things yet to see. Look at the portion of the Milky Way shown in Fig. 129. In the middle of the picture is a great, dark, snake-like object, and a little lower and, the left are two other dark areas. The stars seem to be absent from these spaces, and, of course, We should like to know the reason why.

Two explanations have been offered. There are multitudes of stars all round, with nebulous patches among them, but it may be that these dark spaces are actual opening through the star clouds, and that there are no stars in these directions at all. Few astronomers accept this view. They think it more probable that there is some kind of matter – dust or fine, dark matter of some sort – far out in the depths of space which happens to be between us and these parts of the Milky Way and prevents us from seeing them. The few odd stars which are seen in the dark areas are probably separate individual stars which are in front of the dark matter – that is, they are somewhat nearer to us than the dark matter is.

Here is another photograph of the same sort (Fig. 130). In it is a remarkable mixture of dark areas, star clouds, and bright nebulous patches. The little portion of the sky shown in this picture is in the thickest part of the Milky Way. What a wealth of strange objects there are far out in space!

Another strange combination is to be found in Fig. 131. Here we have a large, bright nebulous area, many smaller bright patches, and several dark areas of different shapes. The large, bright object is known as the "Pelican" Nebula. It is found in the constellation Cygnus, about 2° east of the bright star Deneb (or Arided).

Perhaps the most remarkable object of all is that in the next photograph (Fig. 132). In the sky it is found just south of the most easterly star in the Belt of Orion. It is known as the "Dark Bay" Nebula. This great dark mass completely hides the stars behind it, and you notice that its edge is faintly illuminated by stars in the neighbourhood. Above and to the left is a bright nebula.

FIG. 129. DARK MARKINGS IN THE MILKY WAY

The dark portions are believed to be due to great masses of dust ("cosmic dust") or other semi-opaque matter which hides the stars behind it. If a cloud of such matter is near a bright star it is illuminated by the starlight and is faintly seen. From this photograph it would appear that much of this dusty material is scattered through the part of space represented in this picture. These markings are in the constellation Cygnus. The photograph was taken with the 100-inch telecsope, with an exposure of two hours and forty-five minutes.

Photograph by Duncan, Mount Wilson Observatory

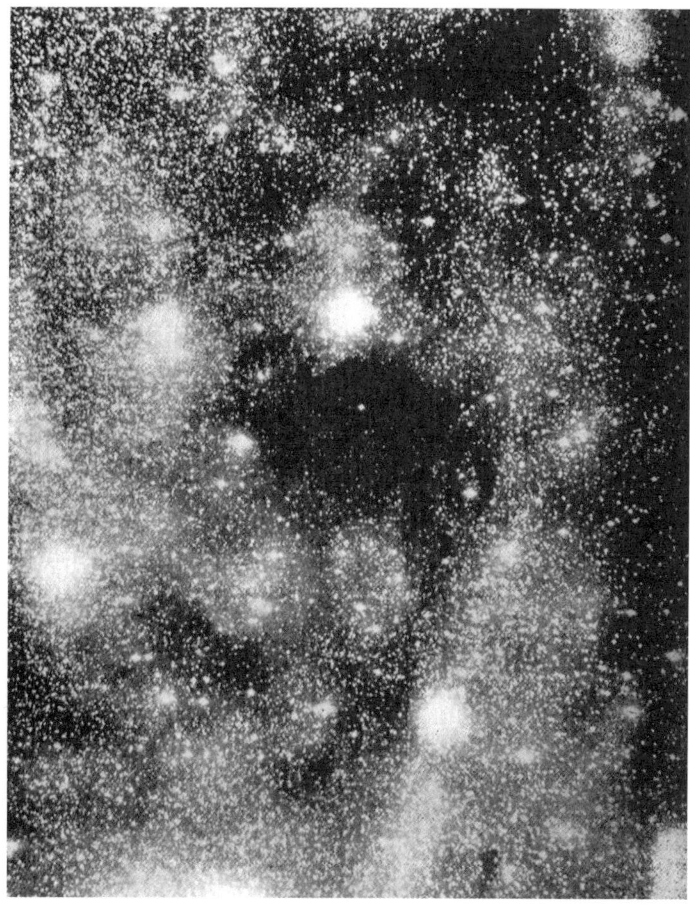

FIG. 130. A DARK HOLE IN THE MILKY WAY
In the telescope this spot appears perfectly black. The stars apparently seen in it are really in front of the dark matter which produces it. The photograph was taken with the 100-inch telescope, with un exposure of four hours.
Photograph by Duncan, Mount Wilson Observatory

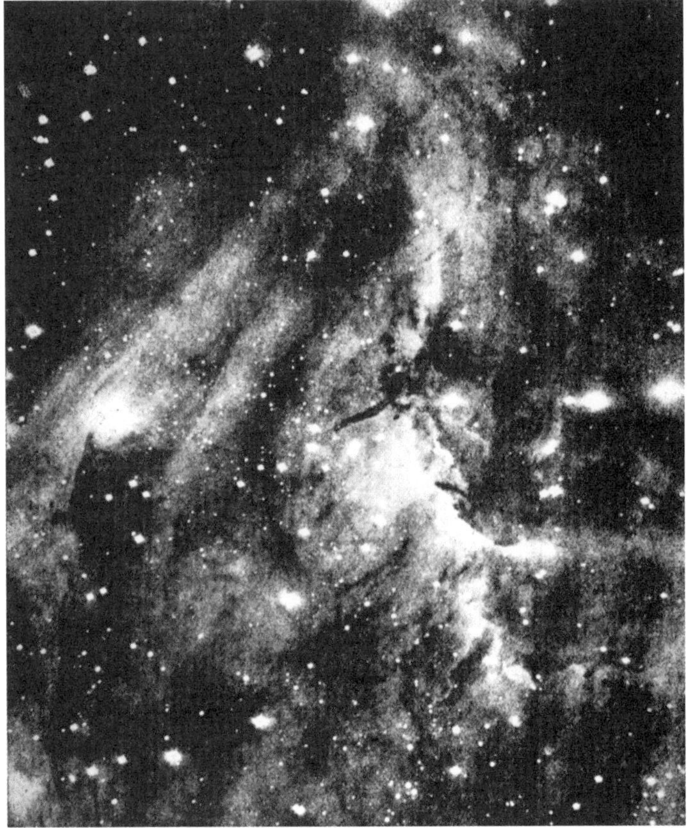

FIG. 131. THE "PELICAN" NEBULA

The portion of the sky shown in this photograph is about 3° east of the star Deneb in Cygnus. The bright nebulosity is evidently dusty material surrounding the stars which lighten it up. The dark streaks and spots, though they look small, are due to immense opaque clouds of dust which hide the bright background from us. The bright patches are probably due to nebulous mutter illuminated by stars within it. Puzzle: find the pelican. This photograph was taken with the 100-inch telescope, and the exposure was four hours and forty-five minutes.

Photograph by Duncan, Mount Wilson Observatory

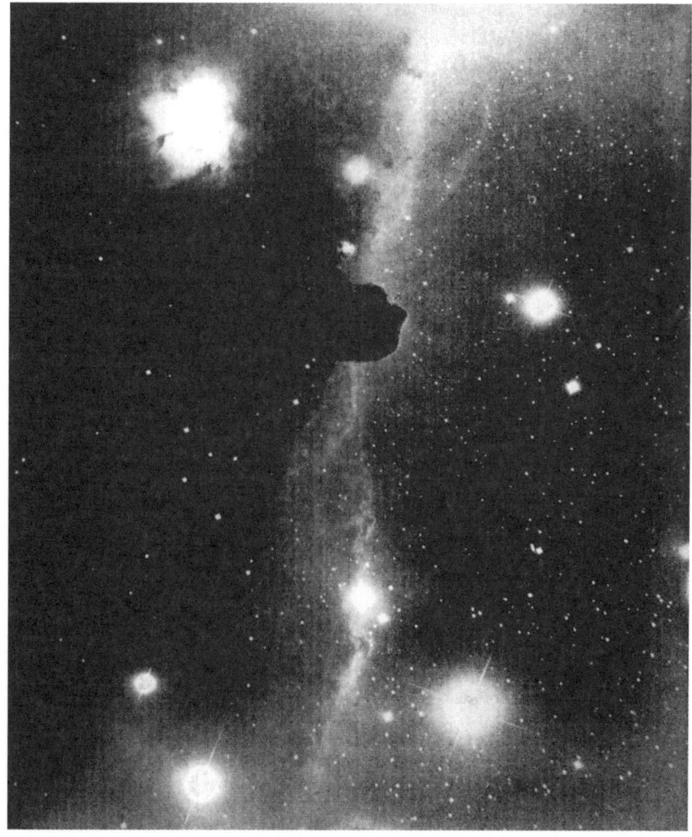

FIG. 132. THE "DARK BAY" NEBULA

This remarkable object is just south of Orion, which is the most easterly star in the belt of Orion. Note the faint nebulous streak, or band, running down the middle of the photograph, and dividing it into two parts. In the right portion numerous faint stars are visible; in the left only a few stars can be seen. Evidently a great dark cloud cuts off the view of the faint stars on the left, those stars which we see being in front of the cloud and seen against it as a dark background.

The photograph was taken with the 100-inch telescope, with an exposure of three hours.

Photograph by Duncan, Mount Wilson observatory

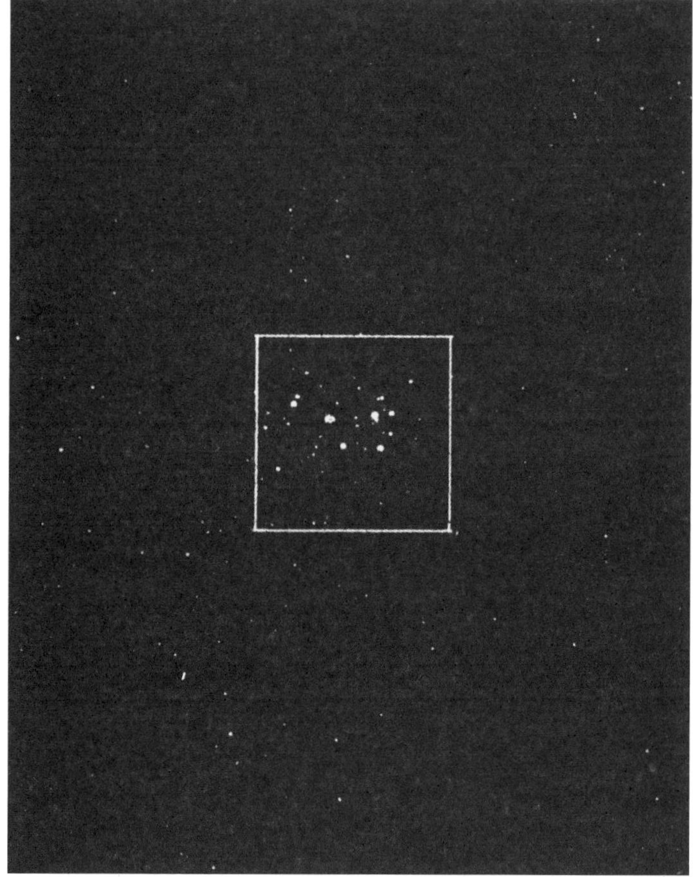

FIG. 133. THE PELADES – AN OPEN CLUSTER
Six stars are easily seen on a dark night. In a 3-inch telescope about a hundred are visible, and a photograph reveals many more. The brightest star is called Alcyone. This group is about 300 light-years distant. Stars down to the thirteenth magnitude are shown in this photograph. The part in the frame is in Fig. 134.
Photograph by Wallace, Yerkes Observatory

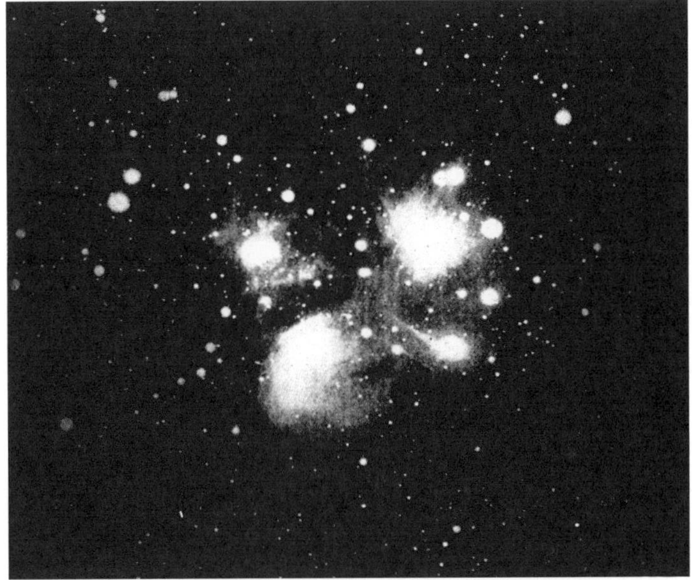

FIG. 134. THE PLEIADES, ENVELOPED IN NEBULOUS MATTER

A long-exposure photograph reveals the fact that the stars in this group are enveloped in nebulous matter. By means of the spectroscope we find that the light from the nebula is just the same as that from the star within it, and hence it is concluded that the matter surrounding the star is probably a dark cloud which is made luminous by the light from the star within it.

Photograph by Bernard, Mount Wilson Observatory

Thus we have evidence of the existence of nebulous matter distributed widely throughout space, mostly in a very rare condition, though it is denser in some localities. This matter is sometimes referred to as 'cosmic dust' or as 'world stuff'. Now the sun, with its attendant planets and their satellites, is travelling through space toward the constellation Lyra at the rate of about twelve miles per second, and it is quite likely that during the past ages our system has passed, at different times, through some of these nebulous masses. This would undoubtedly cause variations in the amount of heat received by the earth from the sun and, as a consequence, changes in the temperature

FIG. 135. PRÆSEPE (THE BEHIVE)
In this cluster, which is the constellation Cancer, the stars are closer together than in the Pleiades, and they seem to be seperate suns quite free from nebulous surroundings.
Photograph by Barnard, Yerkes Observatory

of the earth. These changes may have produced the glacial periods shown in the records of the rocks.

Clusters of Stars

We have already learned of the little group of stars called the Pleiades on the neck of Taurus (the Bull). In Fig. 133 is a photograph of it and the surrounding stars. With the naked eye six stars can easily be seen, while better eyes can see eight.

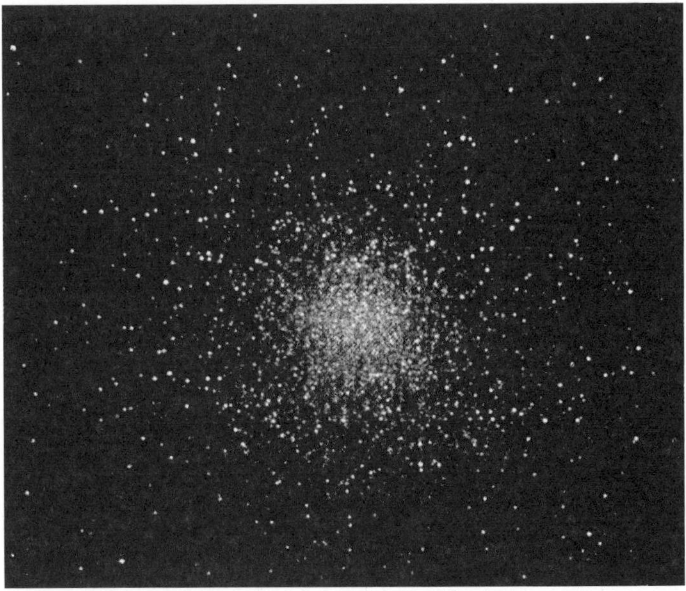

FIG. 136. THE GREAT CLUSTERS IN HERCULES

Here the stars appear to be gathered in a great globe. The stars are all very faint, but the individual ones can be seen in this photograph. Many fainter ones are dimly seen in the background. These globular clusters are at immense distances, ranging from 21,000 to 230,000 light years. This photograph was taken with the 72-inch telescope of the Dominion Astrophysical Observatory, California, with an exposure of one hour.

Photograph by Plaskett, Victoria, British Columbia

Indeed, very sharp eyes have seen fourteen, but a photograph shows about a thousand.

On giving a very long exposure the photograph (Fig.134) brings out something else. It reveals the fact that these stars are simply buried in nebulous matter and, indeed, that all the space near them is full of it.

There are so many things in the sky which the eye alone cannot detect – probably more than we can see even with assistance.

A cluster of stars not quite so open as the Pleiades is shown in Fig. 135. It is called Præsepe (Beehive), and its position in Cancer (the Crab) is shown in the star maps (Figs. 104 and 106). You can recognize it in the sky with the naked eye, but you cannot pick out the separate stars.

The finest clusters are those in which the stars appear to be packed together into a globular form. About seventy such globular clusters are known. The most perfect one in the northern hemisphere is shown in Fig. 136. lt is found on the side of the Flowerpot in the constellation Hercules, and was referred to when we were looking at the summer stars (p. 147). This photograph was taken at the great observatory at Victoria, British Columbia, and the exposure was one hour.

At the centre the stars are close together, but yet you can pick out the individual ones. Farther out many fainter ones are seen, and if a long exposure is given still more come into view.

This is well shown in the frontispiece. For this photograph an exposure of eleven hours was given, spread over three successive nights. The camera was covered during the daytime and then uncovered again at night. At the centre the star images merge into one another; but look at the multitudes farther out.

How many stars do you think there are in this globular cluster? An attempt has been made to count them. Having marked off a certain portion of the cluster, those in it were counted, and then the total number was estimated. We are told that there are upward of 50,000! In the photograph they look close together, but actually each is at least a million miles away from its nearest neighbour – indeed, probably a million million miles, as the distance of the cluster is estimated at 36,000 light-years. Remember, also, that each is a sun like our own, perhaps with planets revolving about it. Can you think of a more wonderful object?

What are the Stars made of?

Finally, consider for a moment what is in the stars. Though they are at enormous distances, they are continually sending forth messages which are carried by means of their light-waves, and with the assistance of the spectroscope the astronomer can interpret what they say.

We learn that these millions of suns, scattered throughout space in all directions, are composed of iron, hydrogen, sodium, carbon, and other substances which are known on the earth. All the heavenly bodies are built up from the same materials. There is a wonderful unity, or oneness, in the entire universe, and the thought comes to us that it is constructed according to a definite, intelligent plan, and we feel that there is an Infinite Mind behind it and controlling it.

When with the mind's eye we look out upon the planets revolving about the sun, and the satellites revolving about the planets, each following its appointed path and at the same time rotating on its axis in its own definite period; and then, looking farther away, behold the hosts of the stars and the nebulæ, almost infinite in number, in distance, and in size, but all made of the very substances which are familiar to us on the earth, some slight indication of the greatness of the universe is revealed to us, and we are surely ready to agree with the Psalmist when he exclaims, "The heavens declare the glory of God; and the firmament sheweth his handywork."

APPENDIX

SOME INTERESTING ASTRONOMICAL FACTS

I. THE SOLAR SYSTEM

1. The Sun and the Moon

Name	Diameter in Miles	Mass (Earth = 1)	Volume (Earth = 1)	Density (Earth = 1)	Rotation on Axis	Gravity at Surface (Earth = 1)
Sun	866,000	332,000	1,300,000	1.39	24⅔ days	27
Moon	2,160	1/81.5	1/49	3.39	27⅓ days	⅙

2. The Planets

Name	Mean Distance in Millions of Miles[1]	Revolution Period	Average Speed in Orbit Miles per Hr.	Diameter in Miles	Mass (Earth = 1)	Rotation on Axis
Mercury	36	88 days	30	3,100	1/24	88 days
Venus	67	225 "	22	7,700	0.81	Undetermined
Earth	93	365¼ "	18½	7,918	1.00	23 h. 56 m. 4.09 s.[2]
Mars	141½	687 "	15	4,215	0.11	24 h. 37 m. 22.6 s.
Jupiter	483	11.86 years	8	86,720	317	9 h. 55 m.
Saturn	886	29.5 "	6	71,500	95	10 h. 14 m.
Uranus	1.782	84 "	4¼	32,400	14.6	10¾ h.
Neptune	2,793	164.8 "	3⅓	31,000	17.2	19 h. (?)

[1] The mean distance of a planet is one-half the longest diameter of its elliptical orbit.

[2] The true rotation period of the earth is not the ordinary day, which is the interval from one noon to the next. The sun is apprently moving eastward in the sky, and the rotation must be determined from the stars.

3. The Number of Satellites of each Planet

Planet	Number of Satellites	Planet	Number of Satellites
Mercury	0	Jupiter	9
Venus	0	Saturn	9
Earth	1	Uranus	4
Mars	2	Neptune	1

II. THE STARS

1. The Twenty Brightest Stars

Arranged in order of brightness

Name of Star	Magnitude	Velocity Miles per second	Distance in Light years	Light (Sun = 1)
a Canis Majoris (Sirius)	- 1.58	12	9	26
a Carinæ (Canopus) (S)	- 0.86	18	650	80,000
a Centauri (S)	0.06	20	4	1.3
a Lyræ (Vega)	0.14	12	26	50
a Aurigæ (Capella)	0.21	26	43	150
a Boötis (Arcturus)	0.24	84	41	100
ß Orionis (Rigel)	0.34	14	543	17,000
a Canis Minoris (Procyon)	0.48	12	10	6
a Eridani (Achernar) (S)	0.60	6	67	200
ß Centauri (S)	0.86	11	270	3,100
a Aquilæ (Altair)	0.89	22	16	9
a Orionis (Betelgeuse)	0.92	14	192	1,200
a Crucis (S)	1.05	11	210	1,650
a Tauri (Aldebaran)	1.06	36	57	90
ß Geminorum (Pollux)	1.21	18	32	28
a Virginis (Spica)	1.21	17	210	1,500
a Scorpii (Antares)	1.22	11	330	3,400
a Piscis Australis (Fomalhaut)	1.29	9	24	13
a Cygni (Deneb)	1.33	12	650	10,000
a Leonis (Regulus)	1.34	11	56	70

The names of the stars in Latin are given first. *a* Canis Majoris is Alpha of the Great Dog, a Carinæ is Alpha of the Keel (of the ship of the Argonauts), and so on.

These twenty stars are usually said to be "of the first magnitude," although they actually differ greatly in brightness.

In the last column is given the relative brightness of the star and the sun, supposing them to be placed at the same distance from the observer.

The letter S indicates that the star is in the southern hemisphere.

2. Star Diameters which have been measured

Star	Diameter in Miles	Star	Diameter in Miles
a Boötis (Arcturus)	23,000,000	ß Pegasi (Scheat)	35,000,000
a Tauri (Aldebaran)	33,000,000	a Herculis (Ras Algethi)	350,000,000
a Orionis (Betelgeuse)	240,000,000	o Ceti (Mira)	260,000,000
a Scorpii (Antares)	400,000,000		

3. Surface Temperatures of the Stars

Colour of Star, with Example	Temperature (C.)	Colour of Star, with Example	Temperature (C.)
Blue-white (Bellatrix)	23,000°	Yellow (Capella)	5,600°
White (Sirius)	11,000°	Orange (Arcturus)	4,200°
Yellowish white (Canopus)	7,400°	Red (Betelgeuse)	3,000°
		Deep red (stars all faint)	2,600°

The temperature is in Centigrade degrees; to change to Fahrenheit multiply by 1.6. The temperature is determined from the spectrum of the star, not simply by its colour. The temperature at the centre of a star is very much higher.

4. Number of Stars of various Magnitudes

Magnitude	Number	Magnitude	Number
1	20	7	15,500
2	57	8	45,000
3	189	9	123,000
4	514	10	330,000
5	1,820	15	27,000,000
6	5,500	20	550,000,000

5. the Velocity of Light

Light travels in 1 second 186,000 (more accurately 186,284) miles.
Light travels 1 minute 11,000,000 miles (nearly)
Light travels 1 year 6,000,000,000,000 miles (nearly)
Thus 1 light year = 6 million million miles

The Life-giving Sun

This scene appears on the golden back of the throne found in the tomb of Tut-ankh-Amen. The king is seated on a cushioned throne, while before him stands his youthful queen. In her left hand she holds a jar of perfume, or ointment, which she is gently applying to the king's shoulder. From above the sun sends down his life-giving rays, each ending in a hand for bestowing gifts. The hands before the king and queen hold the, the symbol of life, which is being offered to them. Behind the king are his two names, Tut-ankh-Amen and Kheperu-neb-Ra. The inscription behind the queen reads, "The Lady of the Two Lands, Ankh-s-p-Aten, brings much fine oil to anoint the King's crown. May it give life eternal and strength unending!"

Drawn from a photograph

INDEX

PUBLISHER'S NOTE

Astronomy is one of the oldest branches of science, fascinating humans from the earliest times.

Huge advances have been made since Clarence Augustus Chant's acclaimed work *Our Wonderful Universe* was first published in 1928. We have sent humans into space and walked on the Moon. Spacecraft have landed on Mars, and the International Space Station, a joint project among five space agencies, has been continuously occupied by humans since November 2000. We are using telescopes and satellites to observe the skies, studying planets, moons, stars, galaxies, and comets, as well as supernovae explosions, gamma ray bursts, and cosmic microwave background radiation. Today's and tomorrow's challenges reach ever further, with key questions such as is there other life in the Universe, and what is the nature of dark matter, and what is the ultimate fate of the Universe?

Astronomy is one of the few sciences where amateurs can still play an important active role, especially in the discovery and observation of variable stars, tracking asteroids and discovering transient objects, such as comets and novae.

This book was written by Chant to excite the wonder of young people and inspire their imaginations. Its purpose and approach is as relevant today, and we hope that readers will enjoy the way in which Chant leads us on his journey of discoveries of the Universe.

★★★

We are grateful to the Royal Astronomical Society of Canada for their help and advice on this new edition.

The publisher would also like to thank Dot Dot Dot for her intuition in bringing the original edition of *Our Wonderful Universe* to our attention, in the hope that we too would fall for its charm and delightful content. She was absolutely right, and we hope that she will be pleased with the finished results.

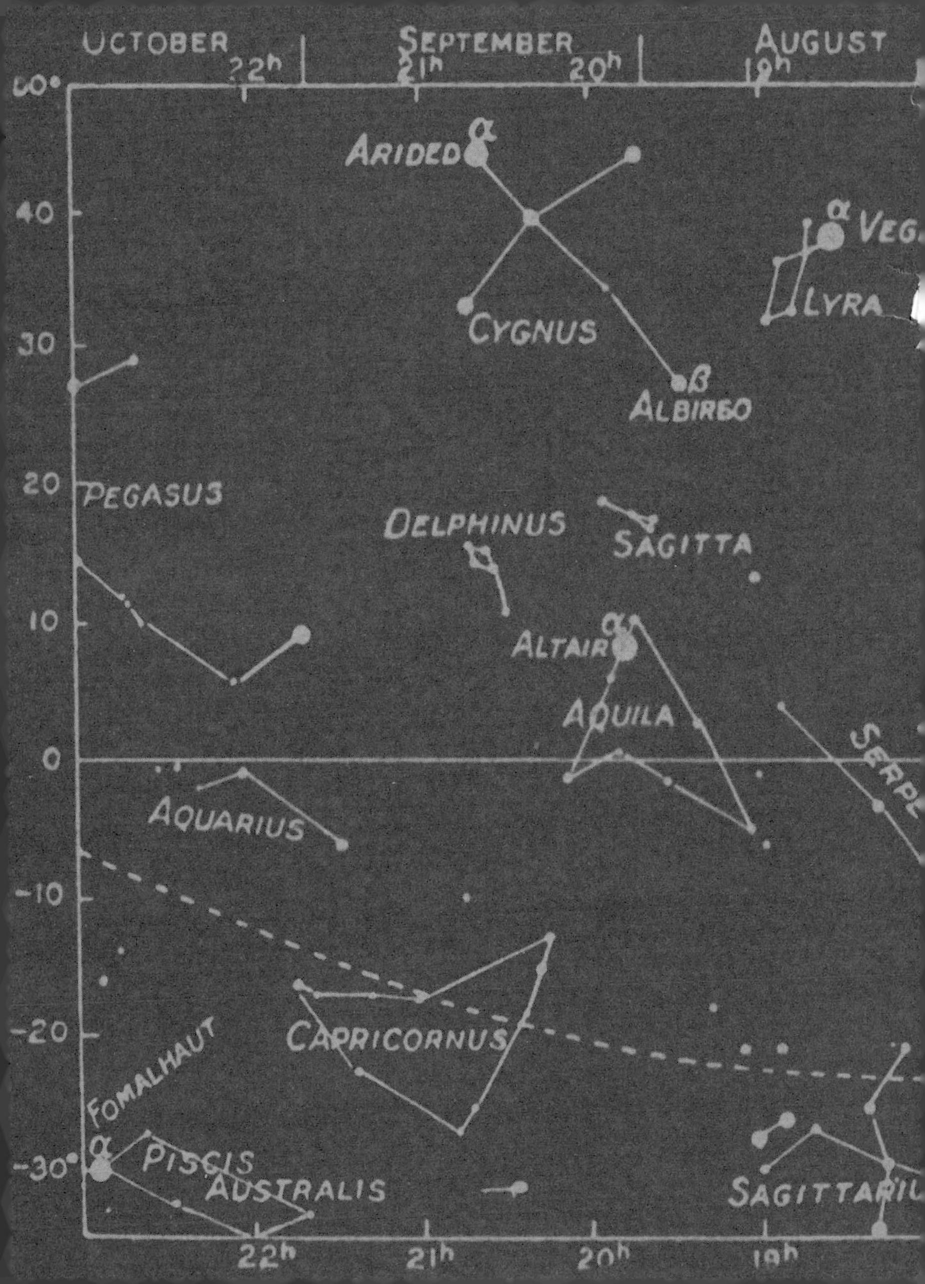

UCTOBER
SEPTEMBER
AUGUST
22h
21h
20h
19h
60°
40
30
20
10
0
-10
-20
-30°
ARIDED
α
CYGNUS
β
ALBIRGO
VEG
LYRA
PEGASUS
DELPHINUS
SAGITTA
ALTAIR
AQUILA
SERP
AQUARIUS
CAPRICORNUS
FOMALHAUT
PISCIS
AUSTRALIS
SAGITTARIU
22h
21h
20h
19h